Advances in Soil Science

Advances in Soil Science

B.A. Stewart, Editor

Advances in Soil Science

Volume 8

Edited by B.A. Stewart

With Contributions by
A.R. Bertrand, J.C. Day, R.J. Gibbs, P.M. Huang
R.E. Meyer, A.S.P. Murthy, R.I. Papendick,
J.B. Reid, and J.L. Steiner

With 52 Illustrations

Springer-Verlag
New York Berlin Heidelberg
London Paris Tokyo

B.A. Stewart
USDA Conservation and Production Research Laboratory
Bushland, Texas 79012, USA

ISSN: 0176-9340

Typeset by Asco Trade Typesetting Ltd., Hong Kong.

9 8 7 6 5 4 3 2 1

ISBN-13: 978-1-4613-8773-2   e-ISBN-13: 978-1-4613-8771-8
DOI: 10.1007/978-1-4613-8771-8

Preface

Soil is formed from physical and chemical weathering of rocks—processes described historically because they involve eons of time—by glaciation and by wind and water transport of soil materials, later deposited in deltas and loessial planes. Soil undergoes further transformations over time and provides a habitat for biological life and a base for the development of civilizations. Soil is dynamic—always changing as a result of the forces of nature and particularly by the influences of man. The soil has been studied as long as history has been documented. W.H. Gardner, in his review "Early Soil Physics into the Mid-20th Century," published in Volume 4 of this series, told of writings on clay tablets, dating to about 1700 BC. Those writings gave specific instructions on cultivating the soil and seeding crops. Numerous references to soil are found in historical writings, such as Aristotle (384–322 BC), Theophrastus (372–286 BC), Cato the Elder (234–149 BC), and Varro (116–27 BC). Some of the earliest historical references to soil 3,000 or more years ago have to do with the erosional forces of wind and water.

The study of soils today has taken on increased importance because a rapidly expanding population is placing demands on the soil never before experienced. Soil scientists have professionally divided themselves into separate disciplines—physics, chemistry, microbiology, mineralogy, genesis, and the like. Studies range from very basic to very applied and to literally every corner of the earth, and of the moon as well.

This series, *Advances in Soil Science*, was established to provide a forum for leading scientists to analyze and summarize the available scientific information on a subject, assessing its importance and identifying additional research needs. It is not the purpose of the series to report new research findings because there are many excellent scientific journals for that need. Communications in scientific journals, however, are generally restricted to short and technical presentations. Therefore, *Advances in Soil Science* fills a gap between the scientific journals and the comprehensive reference books in which scientists can delve into a particular subject relating to soil science. Contributors are asked in particular to develop and identify principles that have practical applications to both developing and developed agricultures.

Advances in Soil Science was formulated to be international in scope and to cover all subjects relating to soil science. This volume certainly fulfills those goals in that it contains contributions from Canada, the United States, India, and New Zealand and presents information ranging from very basic studies on soil aluminum to general discussions about crop production constraints in developing countries. These contributions present valuable information on a diversity of topics and serve as an excellent source of references. Although we consider our audience to be primarily scientists and students of soil science, the series provides technical information for anyone interested in our natural resources and man's influence on these resources.

The ultimate aim of the series is to stimulate action: action to determine where there are arable soils, action to develop technology for more efficient crop production on these soils, action to reduce the risk of degrading these soil resources, and action to determine on which soils our research efforts should be concentrated. Research in the future will focus on systems that are resource efficient and environmentally sound. The need to optimize crop production while conserving the resource base has never been greater.

The quick acceptance of *Advances in Soil Science* by both authors and readers has been very gratifying and confirms our perception that a need did exist for a medium to publish soil science reviews. I want to thank the authors for their excellent contributions and cooperation. I also want to thank members of the Editorial Board for their assistance in selecting such competent authors, and the Springer-Verlag staff for their kind assistance. Lastly, and most importantly, I want to thank the readers for their acceptance and use of *Advances in Soil Science.*

B.A. Stewart

Contents

Contributors

A.R. BERTRAND, Consultant, 1547 Montpelier, Petersburg, Virginia 23805, USA

J.C. DAY, USDA Economic Research Service, Washington, DC 20250, USA

R.J. GIBBS, Soil Science Unit, University College of Wales, Penglais, Aberystwyth SY23 3DE, United Kingdom

P.M. HUANG, Saskatchewan Institute of Pedology, University of Saskatchewan, Saskatoon, Saskatchewan S7N 0W0, Canada

R.E. MEYER, US Agency for International Development, Washington, DC 20523, USA

A.S.P. MURTHY, Advance Centre for Black Soil Research, University of Agricultural Sciences, Dharwad Campus, Dharwad-580005, Karnataka, India; present address: University of Agricultural Sciences, Regional Research Station, V.C. Farm, Mandya 571 405, India

R.I. PAPENDICK, USDA Agricultural Research Service, Washington State University, Pullman, Washington 99164, USA

J.B. REID, Ministry of Agriculture and Fisheries, Batchelar Agriculture Centre, Palmerston North, New Zealand

J.L. STEINER, USDA Agricultural Research Service, Bushland, Texas 79012, USA

Ionic Factors Affecting Aluminum Transformations and the Impact on Soil and Environmental Sciences*

P.M. Huang

*Publication No. R532, Saskatchewan Institute of Pedology, University of Saskatchewan, Saskatoon, Saskatchewan, Canada.

(*cont.*)

I. Introduction

Aluminum is the most abundant metallic element of minerals in soils and the associated environments. It occurs in a series of Al-bearing minerals (e.g., feldspars, micas, chlorites, vermiculites, smectites, kaolinite, halloysite, and gibbsite). It makes up 81, 82, 25, and 4 g kg^{-1} of igneous, shale, sandstone, and limestone rocks, respectively (Jackson, 1964; Brady, 1974; McLean, 1976).

Analogous to carbon as a coordinator for organic matter, Al ranks in abundance next to silicon as an oxygen coordinator in minerals in terrestrial and aquatic environments. Aluminum can be released from minerals to soil solutions and natural waters through chemical and biochemical weathering reactions. Both inorganic and organic ions are an integral part of the environment (Paul and Huang, 1980; Förstner, 1981; Robert and Berthelin, 1986). They are important weathering agents of primary and secondary minerals. The extent of the release of Al from minerals to the environment has increased with time, population growth, intensification of agriculture, and industrialization (Huang, 1987a). The Al released to soil solutions and natural waters undergoes a series of reactions including hydrolysis, polymerization, complexation, precipitation, and crystallization. These reactions are bound to be very significantly influenced by prevalent ionic factors in the systems.

Transformation products of Al as influenced by ionic factors have an important bearing on pedogenesis and soil acidity and liming (Jackson, 1963a,b; Hsu, 1977; Thomas and Hargrove, 1984), formation (Wang *et al.*, 1986), stabilization (Stevenson, 1982), and turnover (Martin and Haider, 1986) of humic substances and soil aggregation and related soil physical properties (Emerson *et al.*, 1986). Furthermore, these Al components have important effects on the fate and dynamics of nutrients and toxic pollutants in the environment (Huang, 1987a). In addition, certain Al ionic species may pose a threat to ecological balance and human and animal health.

II. Forms and Distribution of Aluminum in Soils and Associated Environments

The total concentration of Al in soils and sediments is in general of the same order of magnitude as that of the earth's crust (McLean, 1976). Aluminum is either octahedrally or tetrahedrally coordinated with oxygen in crystalline and short-range ordered Al-bearing minerals. It also exists as exchangeable Al and hydroxy Al interlayers (Hsu, 1977; Huang and Violante, 1986; Bertsch and Barnhisel, 1987) and as coatings on the edges and external planar surfaces (Huang and Kozak, 1970). Exchangeable Al in soil environments is mainly monomeric and may be dimeric to a lesser extent (Jackson, 1963a; Thomas and Hargrove, 1984). Recent research data indicate that the association of organic carbon with exchangeable Al on clay surfaces merits attention (Goh and Huang, 1984, 1986). The hindrance of organic ligands on the hydrolysis and ploymerization of Al via complexation reactions may partially account for the restriction of exchangeable Al to monomeric and dimeric forms in soil environments.

Aluminum interacts with low-molecular-weight organic acids (Kwong and Huang, 1975, 1977, 1979a, 1981; Violante and Huang, 1984, 1985; Huang and Violante, 1986) and fulvic (FA) and humic (HA) acids (Schnitzer and Kodama, 1977; Kodama and Schnitzer, 1980; Schnitzer, 1986) to form insoluble hydroxy-Al-organic complexes, depending on the molar ratio of organic acid to Al, pH, and nature of organic acids. The reaction of Al with inorganic ions results in the formation of a series of sparingly soluble compounds. The K-taranakite is formed in fertilizer zones in soils through the reaction of concentrated K and acid phosphate solution with Al hydroxides (Taylor and Gurney, 1965). If the ionic activities of K^+, Al^{3+}, and SO_4^{2-} are sufficiently high, the K-alunite could be formed in acid soils (Adams and Rawajfih, 1977).

Substitution of Al for Fe in the structure of Fe(III) oxides is known to occur and seems to be widespread in soils (Schwertmann and Taylor, 1982). The degree of Al substitution is related to Al availability in soils and may thus vary with pedogenic environments (Fitzpatrick and Schwertmann, 1982).

In soil solutions and natural waters, Al ions can form complexes with a series of organic and inorganic ligands (Campbell *et al.*, 1985; Huang, 1980, 1987a; Huang and Violante, 1986). However, the information on speciation of dissolved Al is rather limited. Furthermore, the chemical behavior of Al species and their ecological and health effects are even less understood.

III. Fundamental Aqueous Chemistry of Aluminum

A. Hydrolytic Reactions of Aluminum

The Al^{3+} ion, which is released from Al-bearing minerals to soil solutions and natural waters, is octahedrally coordinated with six water molecules and exists as an $Al(H_2O)_6^{3+}$ ion. The $Al(H_2O)_6^{3+}$ is a proton donor, and its K_1 for the following hydrolytic reaction is 1.12×10^{-5} (Cotton and Wilkinson, 1980):

$$Al(H_2O)_6^{3+} = Al(OH)(H_2O)_5^{2+} + H^+$$

Therefore, the $Al(H_2O)_6^{3+}$ ion is moderately acidic. The hydrolysis of $Al(H_2O)_6^{3+}$ in soil solutions and natural waters is thus the rule rather than the exception (Thomas, 1977).

Hydrolysis of $Al(H_2O)_6^{3+}$ ion can proceed through monomeric mechanism (Schofield and Taylor, 1954; Ragland and Coleman, 1960; Frink and Peech, 1963; Hunt, 1963; Frink and Sawhney, 1967; Singh, 1969; Baes and Mesmer, 1976; Stol et al., 1976) and polymeric mechanism (Brosset et al., 1954; Sillen, 1959; Hsu and Rich, 1960; Matijevic et al., 1961; Okura et al., 1962; Jackson, 1963a; Hsu and Bates, 1964b; Aveston, 1965; Fripiat et al., 1965; Hem and Roberson, 1967; Turner, 1969; Turner and Ross, 1970; Smith and Hem, 1972; Baes and Mesmer, 1976; Stol et al., 1976; Hsu, 1977). Monomeric Al ions are not the only Al species, and polymeric Al ions are important hydrolytic products in soil solutions and natural waters.

.1. Monomeric Hydrolysis

There are four mononuclear species of Al; $AlOH^{2+}$, $Al(OH)_2^+$, $Al(OH)_3^0$, and $Al(OH)_4^-$ (Table 1). Of the four mononuclear species, the formation quotients of $AlOH^{2+}$ and $Al(OH)_4^-$ are known most accurately. The formation quotients of the two intermediate species are relatively uncertain (Baes and Mesmer, 1976). May et al. (1979) found the $Al(OH)_3^0$ species to contribute insignificantly to gibbsite solubility and implicitly questioned its existence at the very low total Al concentration studied. The ^{27}Al nuclear magnetic resonance (NMR) spectroscopic investigations of diluted Al standard solutions (Bertsch et al., 1986) indirectly suggests that the $Al(OH)_3^0$ species contributes to the observed line width at half maximum ($V_{1/2}$), although its contribution is relatively insignificant over the narrow pH range investigated. Therefore, the existence of the $Al(OH)_3^0$ species in dilute acidic solution appears to be unresolved.

The solubility behavior of Al hydroxide is consistent with the formation of $Al(OH)_4^-$ in alkaline solutions. The ratio of four OH^- ions per Al^{3+} ions in solution was confirmed by Brosset (1952) and by Mesmer and Baes (1971). The solubility results (Baes and Mesmer, 1976) give:

Table 1. Mononuclear hydrolysis products of Al at 25°C

Species	Medium	Log Q_{1y}[a]
$AlOH^{2+}$	0.1 M $NaClO_4$	−5.30
	0	−4.97
$Al(OH)_2^+$	0.1 M $NaClO_4$	−9.90
$Al(OH)_3$	0.1 M $NaClO_4$	−15.60
$Al(OH)_4^-$	0	−23.00

Source: Baes and Mesmer (1976); reprinted by permission of John Wiley & Sons, Inc.

[a] Q_{1y} is an equilibrium quotient of the hydrolysis of Al.

$$Al^{3+} + 4H_2O = Al(OH)_4^- + 4H^+$$
$$\log K_{14} = -23.0 \pm 0.3$$

At the concentrations below 1.5 M Al, the Raman and infrared spectra (IR) of this species indicate that it is a tetrahedral ion (Moolenaar et al., 1970). They also presented ^{27}Al NMR spectra of these solutions which displayed a narrow resonance line at ≈ 80 ppm downfield from the $Al(H_2O)_6^3$ resonance, with both the line width and chemical shift indicative of tetrahedrally coordinated Al. Studies indicating tetrahedral coordination of the $Al(OH)_4^-$ species were generally conducted at very high base concentrations. From the Raman and IR data of the Al solutions of more moderate base concentrations (1×10^{-6} to 1×10^{-2} M), Carreira et al. (1966) could not resolve if the $Al(OH)_4^-$ species is square planar or octahedral. In more dilute Al solutions of modest base concentrations, in which the formation quotient Q14 was determined, the $Al(OH)_4^-$ species could be a tetragonally distorted octahedral complex that may represent an intermediate in the coordination change.

The $Al(OH)_5^{2-}$ species has been used in calculations of hydrolysis reactions of Al (Lindsay, 1979) and proposed by Maksimova et al. (1967). However, there is no evidence to show its existence.

2. Polymeric Hydrolysis

Aluminum hydrolysis in solution containing relatively low concentrations of Al and low ligand number (ñ) may be described by monomeric hydrolysis mechanism. However, hydrolysis of Al in solutions of either higher Al concentrations or ñ is better described by polymeric hydrolysis mechanism. Generalizations are difficult to make for solutions with Al concentrations and ñ values intermediate to these conditions.

Jander and Winkel (1931) were among the first to suggest the existence of polynuclear Al species based on diffusion coefficient measured in solutions of basic Al salts. Thereafter, many theories on polynuclear Al species have been proposed (Brosset, 1952; Matijevic and Tezak, 1953; Brosset et al., 1954; Sillen, 1959, 1961; Hsu and Rich, 1960; Hsu and Bates, 1964b; Aveston, 1965; Fripiat et al., 1965; Patterson and Tyree, 1973). Hsu and Bates (1964b) proposed a continuous series of polynuclear Al species with the basic unit of the form $[Al_6(OH)_{12}(H_2O)_{12}]^{6+}$. The choice of the hexameric Al species as a basic structural unit is based on the theories of Brosset (1952) and Hsu and Rich (1960). This hexameric ring scheme is similar to the "core link" model of Sillen (1959, 1961) and has been supported by many researchers (Hem and Roberson, 1967; Richburg and Adams, 1970; Smith and Hem, 1972; Stol et al., 1976). With additional results obtained in later studies, Hsu (1977) believes that in the absence of Dowex 50 resin or vermiculite, the hydrolyzed products are likely $[Al_{10}(OH)_{22}]^{8+}$ (double ring) or $[Al_{13}(OH)_{30}]^{9+}$ (triple ring) at NaOH/Al = 2.1 and below, as shown in Figure 1.

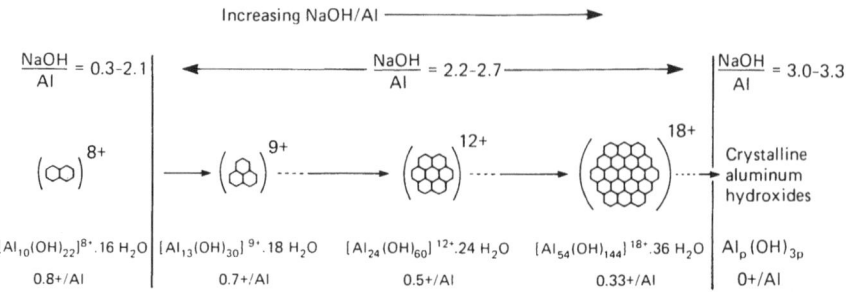

Figure 1. Proposed development of aluminum hydroxides (Hsu, 1977; reprinted with permission of Soil Science Society of America).

Using ultracentrifugation, Aveston (1965) derived average degrees of polymerization (Nw) as a function of ñ and concluded that the $[Al_6(OH)_{15}]^{3+}$ species (Brosset et al., 1954) did not fit the data and the octameric $[Al_8(OH)_{20}]^{4+}$ species (Matijevic et al., 1961) was unlikely. He interpreted his results in terms of the species $Al_2(OH)_2^{4+}$ and $Al_{13}O_4(OH)_{24}^{7+}$. He perferred these two species, because they have been identified in crystalline solids precipitated from hydrolyzed Al solutions. The first occurs as a complex ion in basic aluminum sulfate, $[Al_2(OH)_2(H_2O)_8](SO_4)_2 \cdot 2H_2O$, and in the isomorphous selenate. It consists of two AlO_6 octahedra with a shared edge formed by bridging OH^- ions (Johansson, 1962). The second occurs as the symmetrical ion $Al_{13}O_4(OH)_{24}(H_2O)_{12}^{7+}$ (Figure 2). Aveston (1965) also cited the small-angle X-ray work of Rausch and Bale (1964) who suggested a polymer with

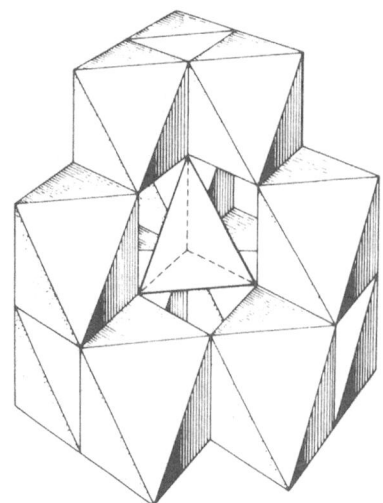

Figure 2. The $[Al_{13}O_4(OH)_{24}(H_2O)_{12}]^{7+}$ ion. The drawing shows how the 12 AlO_6 octahedra are joined together by common edges. The tetrahedra of oxygen atoms in the center of the structure contains one 4-coordinate Al atom (Johansson, 1960; reprinted with permission of *Acta Chemica Scandinavica*).

Table 2. Polynuclear hydrolysis products of Al at 25°C

Medium	$Al_2(OH)_2^{4+}$	$Al_3(OH)_4^{5+}$	$Al_{13}O_4(OH)_{24}^{7+}$
		Log $Q_{xy}{}^a$	
1 M NaClO$_4$	−8.0	−13.47	−104.81

Source: Base and Mesmer (1976); reprinted by permission of John Wiley & Sons, Inc.
^{a} Q_{xy} is an equilibrium quotient of the polynuclear hydrolysis of Al.

a radius of gyration of 0.43 nm, which was assumed to be the Al_{13} polymer, predominated in the solutions investigated.

Baes and Mesmer (1976) conclude that the polynuclear species formed by Al hydrolysis are $Al_2(OH)_2^{4+}$, $Al_3(OH)_4^{5+}$, and $Al_{13}O_4(OH)_{24}^{7+}$ (Table 2). The interpretation of their data indicates acceptable fit with Al_{13} polymer, which is in concurrence with the conclusions of Aveston (1965). Based on ^{27}Al NMR spectroscopic work, Akitt et al. (1972) and Akitt and Farthing (1978, 1981a–d) also proposed the Al_{13} cation. They suggested the predominance of the Al_{13} cation along with small amounts of the dimer and monomeric hexaqua Al ions in similar solutions. Indirect potentiometric studies (Kubota, 1956; Mesmer and Baes, 1971; Bottero et al., 1980) also suggest that the $Al_2(OH)_2(H_2O)_8^{4+}$ species could be a stable polymeric species. More recent evidence from direct NMR spectroscopic investigations (Akitt et al., 1972; Akitt and Farthing, 1981c; Bertsch et al., 1986) further substantiates this contention. The existence of the Al_{13} polynuclear species has been confirmed in crystals of basic Al sulfates precipitated from partially neutralized Al solution (Johansson, 1960) and inferred from small-angle X-ray scattering (Rausch and Bale, 1964), ultracentrifugation (Aveston, 1965), and potentiometric data (Baes and Mesmer, 1976). The charge on the Al_{13} polynuclear species has been reported to range from +3 to +7 (Bottero et al., 1980; Vaughan and Lussier, 1980).

Direct NMR evidence indicates that the Al_{13} polynuclear species exists in partially neutralized Al solutions ranging widely in Al concentration, ñ, and the methods of the synthesis (Bottero et al., 1980; Akitt et al., 1972; Akitt and Farthing, 1981a–d; Bertsch et al., 1986). The resonance peak at ∼ 63 ppm, which is assigned to the Al_{13} polynuclear species, is clearly in the region characteristic of tetrahedrally coordinated Al (Figure 3). Furthermore, the very sharp resonance line with a characteristic line width $V_{1/2}$ of 6–7 Hz suggests both a symmetrical environment with a very weak electric field gradient at its position and general lack of exchange with other Al species. The other 12 octahedrally coordinated Al atoms in the structure display a very broad resonance as a result of their rather asymmetric environment producing a large electric field gradient at their positions. For a given Al concentration, the amount of Al_{13} polynuclear species in partially neutralized Al solutions increases linerally with ñ until high ñ values,

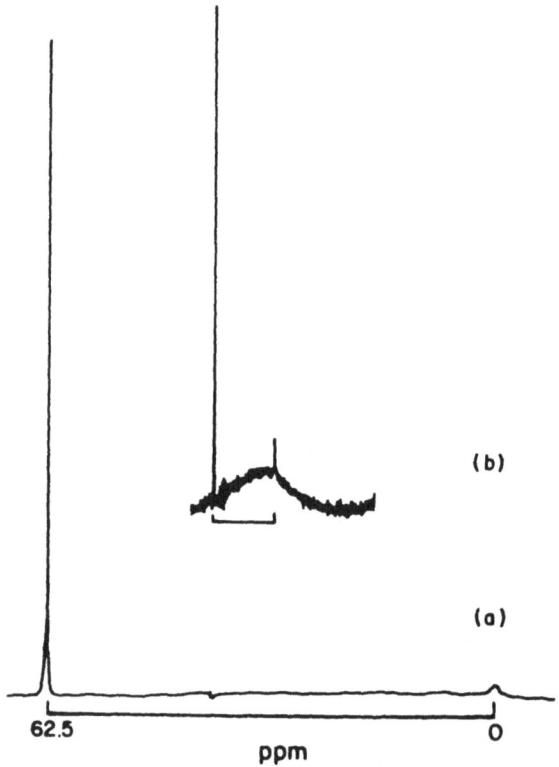

Figure 3. ^{27}Al NMR spectrum of a 0.1-M partially neutralized AlCl$_3$ solution ñ = 2.5 demonstrating the monomeric Al species at 0 ppm and the tetrahedrally coordinated Al of the Al$_{13}$ polymer at 63 ppm downfield from the monomeric Al peak. In the condensed spectrum (b), the broad resonance resulting from the 12 octahedrally coordinated Al atoms in the Al$_{13}$ structure is apparent (Bertsch *et al.*, 1986; reprinted with permission of Soil Science Society of America).

where the Al$_{13}$ polynuclear species begins to decrease (Bottero *et al.*, 1980; Bertsch *et al.*, 1986). The ñ values at which this polynuclear species first forms or begins to decrease depend on many factors including Al concentrations, mode of base addition, base injection rate, and mixing conditions during synthesis (Bottero *et al.*, 1980; Bertsch *et al.*, 1986).

The influence of preparation parameters on the relative amounts of monomeric Al, the Al$_{13}$ polynuclear species, and more polymerized Al species formed during synthesis suggests that the Al$_{13}$ polymers may be formed under inhomogeneous pH conditions at the point of base addition or at the solid solution interface of dissolving carbonate phases (Akitt and Farthing, 1981c; Teagarden *et al.*, 1982; Bertsch *et al.*, 1986). Therefore, it appears that Al(OH)$_4^-$ is a required precursor to the formation of the Al$_{13}$

polynuclear species. The Al_{13} polynuclear species appears to form rapidly when $Al(OH)_4^-$ interacts with 12 octahedrally coordinated Al ions and not to form on aging via polynuclear intermediates (Bottero et al., 1980; Bertsch et al., 1986). The Al_{13} structure could be viewed as being composed of six oxygen-sharing dimeric units which bridge the edges of the central tetrahedron (Akitt and Farthing, 1981c). This species has been shown to be the primary polynuclear species of Al solutions of high ñ (Bottero et al., 1980; Akitt et al., 1972; Akitt and Farthing, 1981a–c; Bertsch et al., 1986).

Recent studies using ^{27}Al NMR, Al-ferron (or hydroxyquinoline) reactions, and sulfate precipitation show that Al polynuclear species in solution gradually change in nature with time, even when the solution maintains similar pH, turbidity, and concentration of mononuclear Al species (Hsu, 1984). These results suggest that the model postulated by Hsu and Bates (1964b) may be oversimplified. The Al_{13} species are present in significant amounts only in freshly prepared solutions of OH/Al molar ratios of 1.5 and above and gradually disappear during aging. Some polynuclear species are not detectable with ^{27}Al NMR spectroscopy. Therefore, no definite conclusion can be drawn at present. Further research on the factors governing the formation and stability of Al_{13} polynuclear species merits close attention in understanding the chemistry of Al pertaining to soil and environmental sciences.

Species other than hexamers and their cyclic derivatives may be present, but their contribution toward the total Al species is much less than that of the hexameric species (Stol et al., 1976). Some species that are deduced from the solid-state structure are shown in Figure 4. Some less stable Al species in solution or transient precipitates may occur in aqueous systems of natural importance. Therefore, the kinetic behavior of these intermediate entities should not be overlooked.

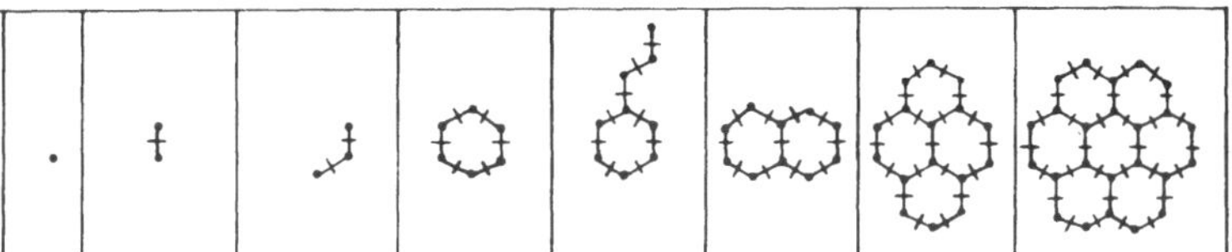

	Al^{3+}	$Al_2(OH)_2^{4+}$	$Al_3(OH)_4^{5+}$	$Al_6(OH)_{12}^{6+}$	$Al_9(OH)_{18}^{9+}$	$Al_{10}(OH)_{22}^{8+}$	$Al_{16}(OH)_{38}^{10+}$	$Al_{24}(OH)_{60}^{12+}$
Average charge/Al	3.0	2.0	1.67	1.0	1.0	0.80	0.63	0.50
OH/Al ratio	0	1.0	1.33	2.0	2.0	2.20	2.38	2.50

Figure 4. Summary of species, which can be deduced from the solid-state structure. The OH/Al ratio refers to the molar ratio in the complexes and not the OH/Al ratio of the system (Stol et al., 1976; reprinted with permission of *Journal of Colloid Interface Science*).

The precise differentiation of polynuclear and mononuclear Al species has been difficult. Indirect methods of differentiation used include conductometry, light scattering, ultracentrifugation, ion exchange, dialysis, and kinetic methods based on the interaction of Al with a complexation agent (Kentamma, 1955; Hem and Roberson, 1967; Hsu and Bates, 1964b; Aveston, 1965; Bersillon et al., 1980; Okura et al., 1962; Turner, 1969, 1971, 1976; Turner and Sulaiman, 1971; Smith, 1971; Smith and Hem, 1972; Patterson and Tyree, 1973; Bertsch et al., 1986; Tsai and Hsu, 1984, 1985). More recently, direct ^{27}Al NMR spectroscopy has provided a unique tool for investigation of the distribution of mononuclear and polynuclear Al species coexisting in solution and the nature of some of the polynuclear species (Akitt et al., 1972; Bottero et al., 1980; Akitt and Farthing, 1981a–d; Bertsch et al., 1986; Buffle et al., 1985; Parthasarthy and Buffle, 1985). ^{27}Al NMR spectroscopy has also cast light on applicability, limitations, and boundary conditions of some indirect differentiation methods. However, even NMR spectroscopy has been unable to provide unequivocal evidence for all the polynuclear Al species present. Continued efforts in differentiating solution Al species would lead to further advancement of knowledge on the chemistry of Al in soils and the associated environments.

B. Precipitation and Crystallization of Hydrolytic Products of Aluminum

Mononuclear and polynuclear Al species can be transformed into colloidal or distinct solid phases. Many of colloidal solid phases initially formed are of very fine particulates that may not settle upon centrifugation and may easily pass membrane filters. The precipitation products of Al can be either short-range ordered materials or crystalline Al hydroxides, depending on solution composition and formation conditions.

In the scheme of polymeric hydrolysis of Al (Figure 1), Hsu (1977) stated that below the OH/Al molar ratio of 3, hydroxyl ions repel each other. As the OH/Al molar ratio reaches 3, net charge per Al is 0; the repulsion becomes negligible, and polynuclear species cluster and form crystals. The transformation of soluble Al species to a solid structure involves the orientation of platelets (the *001* sheets) relative to one another with the help of H bonding and van der Waals forces. During this ordering process, severe constraints exist in the systems.

Aluminum hydroxide may crystallize in three modifications—bayerite, gibbsite, and nordstrandite. In all three of these structures, the a and b crystallographic axes have the same length. Therefore, the (*001*) planes must be identical in structure (Wells, 1962). The differences among three Al hydroxide polymers lie in the stacking of these sheets along the c-crystallographic axis.

Rapid precipitation favors the bayerite structure, and slow crystal growth favors the gibbsite structure (Hsu, 1966). pH also plays an important role in governing the formation of the Al hydroxide polymorphs (Barnhisel and Rich, 1965; Schoen and Roberson, 1970. The acidic medium promotes gibbsite structure, whereas the alkaline medium favors bayerite structure (Barnhisel and Rich, 1965). In acidic medium, Al is present mainly as mononuclear Al^{3+} or $Al(OH)^{2+}$. These Al species have strong hydrolyzing power and therefore favor the formation of gibbsite (Schoen and Roberson, 1970). In alkaline medium, Al is present as $Al(OH)_4^-$. Aluminum is thus precipitated in the form of bayerite. However, gibbsite can be prepared at pH > 12 on an industrial scale, whereas bayerite forms between pH 9 and 12 when a supersaturated sodium aluminate solution is allowed to cool (Hsu, 1977). Furthermore, nordstrandite and gibbsite are more common than bayerite in alkaline soils and bauxite deposits (Lodding, 1961; Keller, 1964; Violante and Jackson, 1979).

The development of aluminum hydroxide polymorphs appears to be related to the rate of precipitation, pH of the system, clay surface, and the nature and concentration of inorganic and organic anions (Hsu and Bates, 1964b; Barnhisel and Rich, 1965; Schoen and Roberson, 1970; Davis and Hill, 1974; Hsu, 1977; Violante and Jackson, 1979, 1981; Violante and Violante, 1980; Violante and Huang, 1985; Huang and Violante, 1986). However, the mechanisms of their formation remain obscure.

Besides Al hydroxide polymorphs, metastable precipitation products are formed during hydrolytic reactions of Al. In several studies where colloidal precipitate was formed during rapid neutralization, it was observed that upon dissolution, polynuclear Al species would result (Turner, 1971; Turner and Sulaiman, 1971). The data obtained from ^{27}Al NMR studies show that heating cloudy solutions resulted in the appearance of Al_{13} polynuclear species in solution, suggesting that the solid phase consisted of discrete Al_{13} units stabilized by a bridging mechanism (Bertsch, 1988). Turner (1971) and Turner and Sulaiman (1971) suggested that some of the initial solid phases formed on rapid neutralization of Al solutions redissolved on aging yielding increase in the polynuclear Al species as estimated by the 8-hydroxy quinoline method. These data appear to indicate the existence of Al_{13} polynuclear units within a colloidal precipitate matrix. ^{27}Al NMR spectroscopic data show that some Al precipitates retained by ultrafilter (13-nm pore size) was composed of Al_{13} polynuclear species, providing further support for the nucleation of this Al species to form discrete units (Bertsch, 1988).

The experimental variables affecting nucleation-precipitation reactions of polynuclear Al species include rates and methods of base addition, pH, localized hydroxyl ion concentrations, nature and concentration of inorganic and organic anions, the sequence of anion introduction to the Al solution, and effects of temperature during synthesis or aging of solutions (Turner, 1971; Turner and Sulaiman, 1971; Mesmer and Baes, 1971; Patterson and Tyree, 1973; Vermeulen et al., 1975; Huang and Violante, 1986).

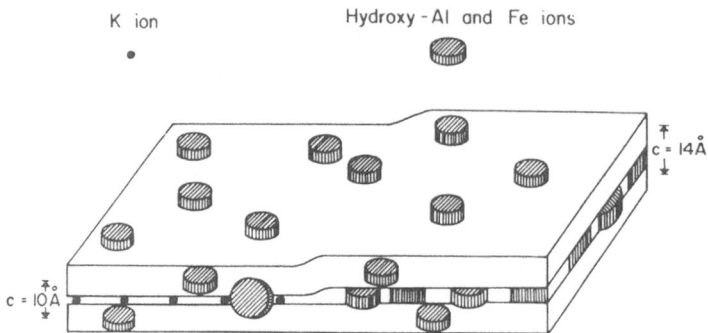

Figure 5. Structural models depicting the hydroxy-Al and Fe coatings on the inter-lamellar and external planar surfaces and edges of mica-vermiculite (Huang, 1980).

IV. Influence of Clay Minerals on Aluminum Transformations

Clay minerals can be considered as solid-state ions and affect Al transformations. Aluminum released to the soil solution through chemical weathering of minerals can be adsorbed by interlayers of 2:1 expansible clay minerals (Jackson, 1963b; Rich, 1968) and external surfaces of phyllosilicates (Huang and Kozak, 1970), as illustrated in Figure 5. The adsorption of hydroxy-Al ions on the external surfaces of phyllosilicates is especially important when the particle size is smaller than 0.2 μm (Table 3). Brosset *et al.* (1954) suggested a six-membered ring structure for the hydrolysis product of Al, having a formula of $Al_6(OH)_{15}^{3+}$ and an OH/Al molar ratio of 2.5. A similar structure was proposed for hydroxy-Al polymers fixed by exchange resins (Hsu and Rich, 1960; Jackson, 1960).

Table 3. Adsorption of hydroxy Al polymers by various-size fractions of muscovite and biotite

	mg of Al adsorbed/g of mineral					
	Muscovite Ratio of hydroxy Al solution to mineral (ml/g)			Biotite Ratio of hydroxy Al solution to mineral (ml/g)		
Size fraction (μm)	25	50	100	25	50	100
5–2	1.4	1.6	3.7	1.7	1.8	3.0
2–0.2	4.0	4.6	5.7	4.5	4.5	5.0
0.2–0.08	13.8	16.5	19.1	13.1	14.8	18.1
< 0.08	13.6	28.4	37.9	13.8	28.3	38.9

Source: Huang and Kozak (1970).

Hydroxy-Al interlayers may consist of these or similar polymers (Jackson, 1962, 1963b; Hsu and Bates, 1964a,b; Hsu, 1977). The existing evidence shows that the OH/Al molar ratio of the nonexchangeable Al occurs in a rather narrow range of about 2.5–2.7. If the OH/Al molar ratio exceeds 2.7, the Al interlayers are metastable (Turner, 1965; Turner and Brydon, 1965, 1967b). The conclusion as to this narrow range of OH/Al molar ratios of nonexchangeable Al is supported by many observations in both natural and synthetic systems (Schwertmann and Jackson, 1964; Singh and Brydon, 1967; De Villiers and Jackson, 1967a,b; Huang and Lee, 1969; Richburg and Adams, 1970; Kirkland and Hajek, 1972).

Smectite with synthetic interlayers resembles natural hydroxy-interlayered smectite in mineralogical and chemical properties. In contrast, this has not been the case for vermiculite. Synthetically prepared hydroxy-interlayered vermiculite often did not remain at a 1.4-nm d-spacing when K-saturated or when subjected to heat treatments. Naturally interlayered vermiculites can be heated to at least 100 or 300°C with little change in the d-spacing, whereas synthetically prepared vermiculite tends to collapse toward a 1.0-nm d-spacing at a lower heat treatment (Barnhisel, 1965). Many factors may be responsible for the low stability of Al interlayers in vermiculite in the synthetic systems. In contrast to a more uniform distribution of hydroxy-Al polymers in smectites, an "atoll" structure is formed in synthetic Al interlayers in vermiculites (Frink, 1965). This concept of two types of structures is supported by the work of Sawhney (1968), Barnhisel (1969), Novak et al., (1971), and Kozak and Huang (1971). The atoll structure results in a steric blocking of exchange sites in the center of the particles and insufficient numbers of props. Therefore, synthetically prepared Al-interlayered vermiculites cannot maintain the d-spacing in the c-axis at 1.4 nm when heated and K-saturated, and the structure collapses toward 1.0 nm. The interlayer space of vermiculites is less than for smectites. The particles of vermiculites are in general larger and likely more rigid than smectites. The charge density of vermiculites is higher than that of smectites, and hydroxy Al polymers may thus deposit more frequently along the crystal edges. The OH/Al molar ratio of Al interlayers in vermiculites may be smaller than in smectites. These factors may contribute to the lower thermal stability of synthetic Al interlayers in vermiculites. In nature, the formation of Al-interlayered vermiculites may proceed over long periods. Therefore, the above-noted factors may be relatively insignificant.

Besides the OH/Al molar ratio, nature of clay surfaces, and particle size and charge density of clays (Barnhisel and Rich, 1963, 1965; Jackson, 1963b; Hsu and Bates, 1964a; Carstea, 1968; Jackson et al., 1973; Hsu, 1977), other factors affecting the stability of synthetic Al interlayers include: (1) nature and amount of hydroxy Al ions (Turner and Brydon, 1965, 1967a; Brydon and Kodama, 1966); (2) anions or salts present during formation of interlayers (Barnhisel and Rich, 1965; Singh and Brydon, 1967); (3) presence of gibbsite in the system (Turner and Brydon, 1967b);

and (4) temperature at which aging is conducted (Carstea, 1968; Singh, 1972). Although optimum conditions for hydroxy-Al interlayer formation in nature are not entirely known, the favorable conditions appear to be as follows: (1) moderately active weathering must be in progress or have occurred to furnish Al ions; and (2) the pH should be moderately acid, about pH 5.0, organic matter content should be low, and there should be frequent wetting and drying of the soils (Rich, 1968). The ability of 2:1 expansible clays to fix polynuclear Al has been proposed as a reason for the absence of gibbsite in temperate soils and is referred to as the "antigibb-site" effect (Jackson, 1963a,b).

In addition to the formation of Al-interlayered clays displaying a stable 1.4-nm d-spacing, it has recently been reported that direct magic-angle spinning NMR evidence substantiates the reasoning that Al_{13} polynuclear species can be deposited in the interlayer space of the 2:1 clay minerals with a stable 1.8-nm d-spacing (Bertsch and Barnhisel, 1987). However, the structural and surface properties of these clay minerals in affecting the formation of Al_{13} interlayers remain obscure.

Little is known about the mechanisms of the polymerization reactions of Al at the mineral water interface and the importance to the formation of polynuclear Al species, particularly in moderately acid environments, where solution Al concentrations are low. Even though Al concentrations in bulk solutions are low, the Al concentrations on the clay surfaces can be several orders of magnitude higher, thus facilitating the formation of poly-nuclear species. Hydrolysis of Al on clay surfaces has been demonstrated (Barnhisel and Bertsch, 1982; Bertsch and Barnhisel, 1987). Polynuclear Al species can form on the surfaces of clay minerals by simply washing a partially Al saturated clay with water; there is some indication that Al_{13} polynuclear species may form in this manner (Perrott, 1981). High concentrations of Al on clay mineral surfaces and the influence of surface charge on hydrolytic reactions of Al are certainly favorable for the formation of polynuclear Al species, even in systems with relatively low concentrations of Al. However, the influence of surface properties of clay minerals on the kind of polynuclear Al species formed at moderate Al surface coverage is not known. The surfaces of clay minerals may serve as sites for the formation of polynuclear Al species in soil solutions and natural waters, since some polynuclear Al species are exchangeable to other cations (Bertsch and Barnhisel, 1987).

V. Influence of Inorganic Ligands on Aluminum Transformations

The nature and concentration of inorganic ligands play an important role in influencing the hydrolysis and polymerization of Al (Marion and Thomas, 1946; Ross and Turner, 1971; De Hek *et al.*, 1978; Nordstrom, 1982;

Violante and Huang, 1984, 1985). The residual positive charge of poly-nuclear Al (Figure 1) must be balanced by the counteranions, giving rise to basic salts.

With anions such as Cl^-, ClO_4^-, and NO_3^- (Hsu, 1966; Ross and Turner, 1971), which do not have strong affinity for Al, the basic salts are highly solube and yield clear solutions (Hsu, 1977). Furthermore, polynuclear Al ions continue to hydrolyze and polymerize into larger units during prolonged aging unless a high concentration of the counteranion is present.

Pseudoboehmite is formed in the presence of a high concentration of indifferent electrolytes such as NaCl (Hsu, 1967; Chesworth, 1972). The data suggest that pseudoboehmite may form in a saline environment. Hsu (1967) also provides evidence showing the gradual transformation of pseudoboehmite to Al hydroxide during aging. The pseudoboehmite formed in a saline environment may persist in nature for a very long time if it is dried before its conversion to Al hydroxide. Materials similar to pseudoboehmite formed in the laboratory have been found in submicroscopic particles in most of the European bauxites (Lippens and Steggerda, 1970). Noncrystalline alumina of boehmite character has been found in certain tropical soils (De Villiers, 1969). Although pseudoboehmite is not identical with the typical boehmite, both of them consist of Al-O-Al linkages. The difference between pseudoboehmite and typical boehmite is a matter of rate rather than the basic nature of the reaction.

In the presence of ligands that have strong affinity for Al, such as sulfate (Hsu, 1973), phosphate (Hsu, 1975), silicic acid (Luciuk and Huang, 1974), carbonate (White and Hem, 1975; Serna et al., 1977; Bardossy and White, 1979), and fluoride (Violante and Huang, 1985), the further hydrolysis of polynuclear Al species is prevented or at least retarded. These counterpolyvalent anions tend to link polynuclear Al species together but in distorted arrangement because of steric reasons. Most of such precipitation products are thus short-range ordered materials and different from Al hydroxide polymorphs. These ligands, with strong affinity for Al, can also promote physical flocculation resulting from reduction in electrostatic repulsion between polynuclear Al ions (Serna et al., 1977; De Hek et al., 1978). Nail et al. (1976a,b) proposed that the initial precipitation products of Al was composed of polynuclear Al species with the gibbsite fragment structure bound together by anions, which was also thought to inhibit further crystallization by restricting growth along edges. The same type of mechanism has been suggested for the interactions of complexing ligands with Al_{13} polynuclear species (Teagarden et al., 1981).

Inorganic ligands that have strong affinity for Al, such as phosphate (Violante and Huang, 1985) and sulfate (Hsu and Bates, 1964b; Violante and Huang, 1984, 1985), promote and stabilize the formation of pseudoboehmites in the optimum range of the ligand/Al molar ratios. Pseudoboehmite has never been found to be formed in the presence of fluoride. If the ligand/Al molar ratios are sufficiently high, these ligands inhibit the

Table 4. Precipitation products of aluminum formed in the presence of inorganic ligands after 5 months of aging at 20°C[a]

Ligand/Al molar ratio (R)	Ligand			
	Sulfate	Silicate	Fluoride	Phosphate
0.005	B[b]	—	B	B
0.01	B	—	—	—
0.02	B, (G)	B	B	G, P
0.05	B, (G)	B, (P)	B, S	P
0.1	B, (G)	B, G, (P)	B, S	P
0.2	B, G	B, P	B, S	A
0.5	B, G	A	A	A
1.0	G, (BP)	A	A	A
3.0	G, P	A	A	A

Source: Violante and Huang (1985); reprinted by permission of Clay Minerals Society.

[a] Initial pH = 8.2.

[b] B = bayerite; G = gibbsite; N = nordstrandite, A = X-ray-amorphous material; P = pseudoboehmite; ( ) = small amount; S = gelatinous, shapeless material detected by transmission electron microscopy.

crystallization of Al hydroxides and oxyhydroxides and promote the formation of noncrystalline precipitation products of Al (Violante and Huang, 1985). The influence of these inorganic ligands on the formation of Al hydroxides and oxyhydroxides is shown in Table 4.

The sequence of inorganic ligand introduction to the Al solution also influences Al transformations. The introduction of complexing ligands to the Al solution prior to neutralization results in the inhibitory influence on the hydrolysis-precipitation reactions of Al. On the other hand, if complexing ligands are interacted with a partially neutralized Al solution, even the addition of F^- results in rapid precipitation of polynuclear Al species, with efficient removal of dissolved F^- (Buffle *et al.*, 1985; Parthasarthy and Buffle, 1985).

Complexation of Al by a series of inorganic ligands may have a considerable influence on the behavior of Al in soil solutions and natural waters (Hem, 1968; Roberson and Hem, 1969; Johnson *et al.*, 1981; Huang and Violante, 1986). The ecological significance of such complexation reactions and the impact on human health need to be investigated.

VI. Influence of Organic Acids on Aluminum Transformations

A. Organic Acids in Soils and Associated Environments

The importance of organic acids in the weathering of minerals was reported in the classical studies by Sprengel (1826), Thenard (1870), Julien (1879), and Bolton (1880, 1882). Organic acids are agents for the mobilization and transport of metals in soils and natural waters (Kaurichev and Nozdrunova, 1961; Schnitzer, 1968; Zunino and Martin, 1977; Förstner, 1981; Stevenson, 1982). The carboxyl and hydroxyl groups are the principal functional groups involved in the reactions of metal ions with organic acids in soil solutions and natural waters. These functional groups are present in both the humic and nonhumic fractions of soil organic components (Schnitzer and Skinner, 1965; Schnitzer and Kodama, 1977; Stevenson, 1982). The significance of the humic fractions, such as fulvic acids (FAs), in complexing with Al in solution has been intensively studied (Schnitzer and Kodama, 1977; Kodama and Schnitzer, 1980). The nonhumic fractions, such as the low-molecular-weight organic acids, are of great importance in the translocation of Al and Fe and pedogenesis (Bloomfield, 1964; Bruckert, 1970a,b; Stevenson and Ardakani, 1972; Stevenson, 1982).

The soluble humic substances comprise two major fractions. The low-molecular-weight materials with fairly high functional group acidity (FAs) are not precipitated at pH 2, whereas the larger humic acids (HAs), with molecular weights of 20,000–100,000, are precipitated. The materials that are so intimately associated with the clays that they are not removed by the basic solution or sequestering agents are known as humins. HAs isolated from a wide range of soils have a carbon content ranging from 53.6% to 58.8% with an average of 56.2%. The FAs have a lower carbon content and also show greater variability in composition. They have higher hydrogen, sulfur, and oxygen, but lower nitrogen contents. This is expressed in the functional groups with FAs having a much higher total acidity, carboxylic acid content, and alcoholic OH than the HAs isolated from the same soil (Schnitzer, 1977).

HAs and FAs behave like linear flexible polyelectrolytes that are readily aggregated at low pH with the aid of hydrogen bonding, van der Waal's interactions, and interactions with the π electron systems of adjacent molecules (Flaig et al., 1975). FAs at pH values above 2–3 occur as elongated fibers and bundles of fibers with a relatively open structure. With increases in pH, the fibers tend to mesh into a woven network, yielding a spongelike structure. HAs show similar structures, but because of the lower solubility in water the same structures are observed over a narrower pH range. A contemporary view of the structure of humic substances has been presented by Hayes and Himes (1986). The composition of humic materials is not identical in all soils. These substances are associated to varying extents (as much as 40–50% of the dry weight) with hydrolyzable carbohydrates and peptide or polypeptide materials. They are virtually bound to soil mineral colloids.

The sources of low-molecular-weight organic acids in soil environments include root exudates, canopy drip, decay of plant and animal residues, and microbial metabolites. The organic acids in soils can be broadly grouped into three categories: the volatile aliphatic acids, the nonvolatile aliphatic acids, and the aromatic acids (Stevenson, 1967). The relative abundance of these three categories of organic acids in soils generally

Table 5. Levels of some common organic acids in soil solutions

Nomenclature of acids	Concentration of acids (M) $\times 10^5$	Reference
Acetic	370–500	Stevenson and Ardakani (1972)
	265–570	Rao and Mikkelsen (1977)
Amino	8–60	Stevenson and Ardakani (1972)
Benzoic	7.5	Whitehead (1964)
Citric	1.4	Bruckert (1970b)
Ferulic	0.1–3.2	Whitehead (1964)
	0.0–0.3	Wang et al. (1967)
Formic	250–435	Stevenson (1967)
Malic + malonic + tartaric	100–400	Stevenson and Ardakani (1972)
Oxalic	6.2	Bruckert (1970b)
p-Coumaric	0.9–4.2	Whitehead (1964)
	0.0–2.2	Wang et al. (1967)
p-Hydroxybenzoic	0.8–3.9	Whitehead (1964)
	0.0–0.9	Wang et al. (1967)
Propionic	19	Walters (1916)
Vanillic	0.7–4.9	Whitehead (1964)
	0.0–0.7	Wang et al. (1967)
Tannic acid–related compounds	5–30	Coulson et al. (1960); Davies (1971); Stevenson (1982)

follows the order volatile aliphatic acids > nonvolatile aliphatic acids > aromatic acids (Wang et al., 1967; Casalicchio and Rossi, 1970). The levels of organic acids in soils are affected by a number of variables. Anaerobic conditions are particularly favorable for the microbial synthesis of organic acids (Desai and Rao, 1957; Sato and Yamane, 1967). Their concentration range in soil solutions is presented in Table 5.

Living plants contribute organic acids to the soil in root exudates and leaf washings. The exudates contain a range of organic acids such as oxalic, citric (Rovira and McDougall, 1967), and uronic acids (Oades, 1978) and a series of amino acids including glutamine, proline, and methionine (Bokhari et al., 1979). Bruckert et al. (1971) identified citric, malic, oxalic, succinic, vanillic, p-hydroxybenzoic, and p-coumaric acids in rain washings of plant leaves. Gomah and Sakhar (1972) observed phenolic acids such as gallic acid among the substances released from plant leaves by leaching with water.

Plant cells contain a range of carboxylic acids, such as citric, quinic, and shikimic acids (Clément, 1977). The amounts present are related to the kind of plants and the soil type on which the plants are growing. These acids are released by autolysis of the death of the plant. As plant tissues decay, they are attacked by the microflorae that synthesize a similar range of metabolites. Therefore, the origin of these compounds in soils cannot be easily distinguished (McKeague et al., 1986). Bacteria flourishing on any medium usually produce large quantities of simple volatile acids such as formic, acetic, propionic, and butyric, whereas the greater part of the acids formed by fungi resembles the range of aliphatic acids present in the tricarboxylic acid cycle that are predominant in the cells of higher plants. Some bacteria such as pseudomonades, able to grow in acid conditions, are also able to produce oxalic acid (Berthelin and Belgy, 1979).

Roots and leaves decomposing in soils generate relatively large amounts of acetic, propionic, and butyric acids (Lynch et al., 1980). Organic acids formed during root decomposition include hexanoic, phenylacetic, succinic, cinnamic, 4-hydroxyphenyl propionic, p-coumaric, and 3,4-dihydroxyphenylpropionic acids. 2-Ketogluconic acid accounted for 20% of the products formed in the rhizosphere of wheat seedlings (Triticum vulgare) (Moghimi et al., 1978) and was probably a bacterial product (Duff and Webley, 1959). A large proportion of the bacteria in newly formed soils on rock surfaces produce this acid, which is particularly effective in dissolving minerals (Duff et al., 1963).

Long-chain fatty acids in the range C_{12}–C_{24} are also found in soils in the lipid fraction. Their nature can be related to the plant species that have been growing in the soils (McKeague et al., 1986).

Phenolic acids in soils are considered to originate not only from plants but also from the decomposition of lignin and microbial synthesis (Flaig, 1982). The phenolic acids derived from lignin include p-hydroxybenzoic, vanillic, syringic, and ferulic acids. Water extracts of barley (Hordeum vulgare) and wheat straws have been found to contain p-hydroxybenzoic, ferulic, and p-coumaric acids (Stevenson, 1967). These phenolic acids are present in small amounts in soils. In the clay loam A horizon of an Alfisol, for example, p-hydroxybenzoic acid predominated in the soil solution (1.4 μM) with smaller amounts of vanillic acid (0.11 μM) and very much smaller amounts of p-coumaric and ferulic acids (Whitehead et al., 1981).

Although low-molecular-weight organic acids are present in relatively low concentrations in the soil solution and normally have only a transitory existence in soil environments, they are constantly added to the soils through natural vegetation, farming, and microbial metabolism. Furthermore, since the time required for soil formation can extend over a period of centuries, the cumulative effect of these organic acids on the transformations of Al in soils and associated environments and the impact on soil-plant relationships, environmental quality, and human and animal health merit attention.

B. Influence of Organic Ligands on Hydrolytic Reactions of Aluminum

Organic components are an integral part of soils and associated environments. Organic acids that are present in soil solutions and natural waters form complexes with Al and thus influence the subsequent transformations. The extent of the influence depends on the nature and concentration of organic acids. The occupation of the coordination sites of Al by organic

Table 6. Percentage distribution of precipitated Al in suspensions at the initial Al concentration of 1.1×10^{-3} M and OH/Al molar ratio of 2.0 as influenced by selected complexing organic acids at the end of the 40-day aging period at room temperature

Organic acid	% Al precipitated ($> 0.025 \mu m$) organic acid concentration, M		
	0	10^{-6}	10^{-4}
p-Hydroxybenzoic acid	36.9	35.2	31.7
Aspartic acid	36.9	34.4	27.5
Tannic acid	36.9	32.0	24.4
Malic acid	36.9	30.6	20.5
Citric acid	36.9	12.4	11.0

Source: Kwong and Huang (1979a); reprinted by permission of Williams & Wilkins.

Table 7. Stability constants of the complexes formed between Al and p-hydroxybenzoic, aspartic, tannic, malic, and citric acids at 25°C

Organic acid	Stability constants of the complexes	
	$\log K_1$	$\log K_2$
p-Hydroxybenzoic acid	1.66	—
Aspartic acid	2.60	—
Tannic acid	3.78	—
Malic acid	5.14	8.52
Citric acid	7.37	13.90

Source: Kwong and Huang (1979a); reprinted by permission of Williams & Wilkins.

ligands disrupts the hydroxyl bridging mechanism indispensable for the polymerization of Al ions (Kwong and Huang, 1975, 1977).

Kwong and Huang (1979a) reported that citric, malic, tannic, aspartic, and p-hydroxybenzoic acids hinder the precipitation of solid-phase hydrolytic products of Al (Table 6). Both the kinds and concentrations of organic acids are important in hampering the precipitation of Al. When organic acids are present at the same concentration, they hamper the precipitation of Al in the order citric acid > malic acid > tannic acid > aspartic acid > p-hydroxybenzoic acid. Among the five organic acids, citric acid gives rise to the most stable complexes with Al, and p-hydroxybenzoic acid has the least affinity for Al, as indicated by the stoichiometric stability constants (Table 7). The affinity of the organic acids for Al (Table 7) coincides with their ability, when present at the same concentration, to hamper the precipitation of Al (Table 6).

The positively charged edges of polynuclear Al ions, in the absence of organic acids, undergo hydrolysis as depicted below:

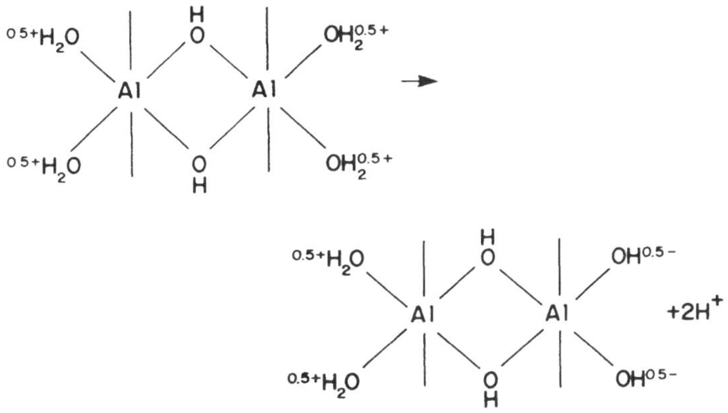

The hydrolysis of these positive edges in the absence of organic acids upon aging results in a decrease in pH (Kwong and Huang, 1979a).

If an organic acid such as citric acid is present in the aqueous solution of Al, the coordination sites of polynuclear Al are occupied by citrate instead of H_2O molecules, imposing a restraint on the hydrolysis of Al as shown below (Kwong and Huang, 1979a).

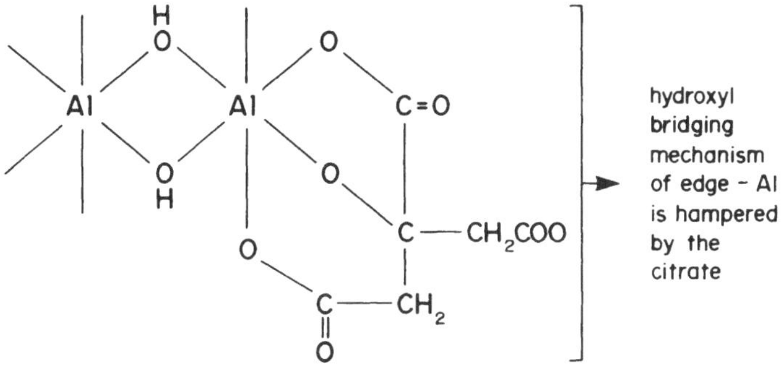

If more organic acids are present in the system, the greater would be the replacement of H_2O molecules and blocking of the coordination sites of Al, and the greater becomes the restraint on the subsequent hydrolysis of polynuclear Al ions. If the concentration of organic acids is held constant, the greater the affinity of the organic acids for Al, the more extensive would be their occupation of the coordination sites of Al, and the more efficient the restraint imposed on the hydrolysis of polynuclear Al ions. Therefore, the precipitation of Al decreases both with the increase of the concentration of organic acids present during the precipitation (Table 6) and with the increase of the stability constants of their complexes with Al (Table 7). The polymerization of Al could occur as interpreted from the broadening of NMR signals of aqueous solutions containing Al chelates of hydroxycarboxylic acids, such as citric and tartaric acids (Toy et al., 1973).

Some coordination sites of Al in Al chelate may be still occupied by hydroxyl groups or H_2O molecules which could undergo hydrolysis to some extent. The occupation of the coordination sites of Al by organic ligands blocks the sites that are indispensable for hydroxyl bridging in the polymerization of Al and its subsequent transformations.

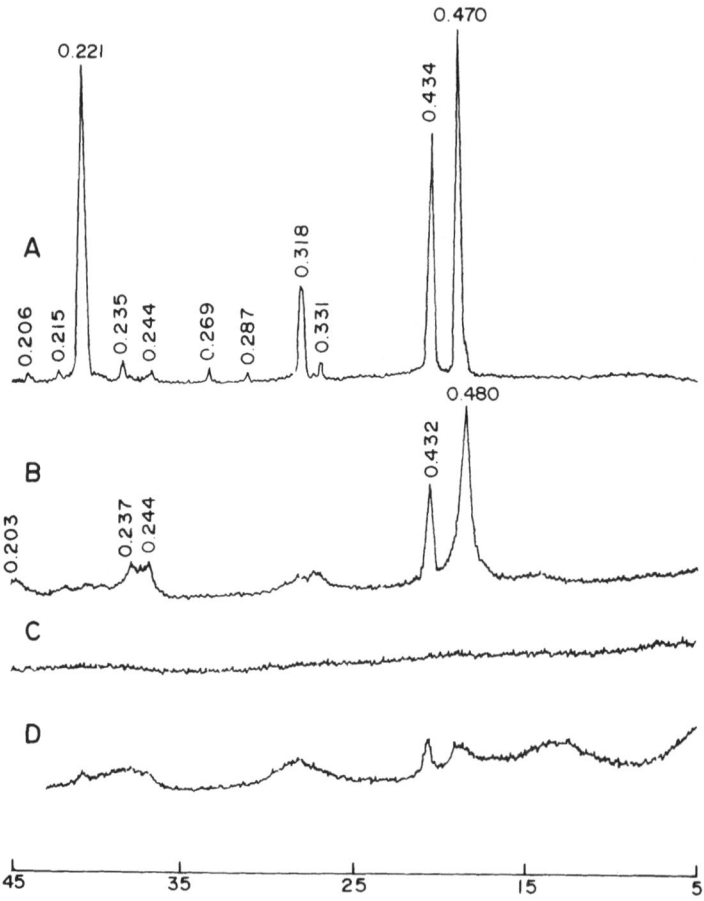

Figure 6. The X-ray diffraction patterns of precipitation products of Al at the initial OH/Al molar ratio of 3 and (A) Al concentration of 1.10×10^{-3} M in the absence of citric acid collected after the 40-day aging at room temperature; (B) Al concentration of 1.10×10^{-3} M in the presence of 10^{-6} M citric acid collected after the 40-day aging at room temperature; (C) Al concentration of 1.10×10^{-3} M in the presence of 10^{-4} M citric acid collected after the 40-day aging at room temperature followed by the 3-day aging at 80°C; and (D) Al concentration of 1.10×10^{-4} M in the presence of 10^{-6} M citric acid collected after the 40-day aging at room temperature (Kwong and Huang, 1975; reprinted with permission of Clay Minerals Society). d-Spacings are in nanometers.

C. Role of Organic Ligands in the Formation of Short-Range Ordered Precipitation Products of Aluminum

The persistence of organic acids in acid soils (Bruckert, 1970a) and the reported occurrence of noncrystalline hydrous oxides of Al under acidic conditions (Mitchell *et al.*, 1964) point toward a possible interference of organic acids in the crystallization of precipitation products of Al. Kwong and Huang (1975) reported that citric acid hampers the crystallization of aluminum hydroxides (Figure 6). The noncrystalline nature of the products is attributed to the occupation of the coordination sites of Al^{3+} ions by citric acid upon dissociation of proton, resulting in a distortion in the arrangement of the hexagonal ring units normally present in crystalline aluminum hydroxides. The structural water of the reaction products is lost more gradually when citric acid concentration is raised from 0 M to 1.4×10^{-4} M (Kwong and Huang, 1977), indicating that a greater structural disorder occurs within the products with increasing levels of citric acid in the systems. This inference from thermal analysis supports the reasoning that citric acid hampers the hydroxyl bridging mechanism in the hydrolytic reaction of Al.

Electron diffraction analyses reveal that the products formed from an initial Al concentration of 1.10×10^{-4} M and OH/Al molar ratio of 1 in the presence of 1.0×10^{-6} M citric acid after the 40-day aging, which are noncrystalline to X-rays, are in effect microcrystalline with reflections at d = 0.110 and 0.181 nm (Kwong and Huang, 1977). The reaction products collected after the 40-day aging from the system with an initial Al concentration of 1.0×10^{-4} M and OH/Al molar ratio of 3 in the presence of a citric acid concentration as high as 1.0×10^{-4} M are also microcrystalline and give reflections at d = 0.321, 0.275, 0.198, 0.166, and 0.142 nm. The microcrystalline nature of the products from the d-values obtained by electron diffraction analyses are still obscure. The electron diffraction data reveal that at the concentrations used, citric acid hampers the formation of even microcrystalline bayerite and gibbsite. They also indicate that the masking effect of noncrystalline products on crystalline cores is not the reason why the solid-phase reaction products formed in the presence of citric acid are noncrystalline to X-rays.

Organic acids, which are common in soil solutions and natural waters, vary in their chemical composition, structure, nature of functional groups, size, and basicity. Therefore, they vary in their ability to perturb the hydrolytic reactions of Al (Kwong and Huang, 1979a). In systems at the initial Al concentration of 1.1×1.0^{-3} M, OH/Al molar ratio of 3 and citric acid concentration of 10^{-6} M, the complexation of Al by citrate, which has the strongest affinity for Al (Table 7), is sufficient to cause the hydrolytic precipitation products formed to be noncrystalline to X-rays after the 1-day aging (Figure 7E). *p*-Hydroxybenzoic, aspartic, and malic acids, which have weaker affinity for Al than citric acid (Table 7), do not complex Al

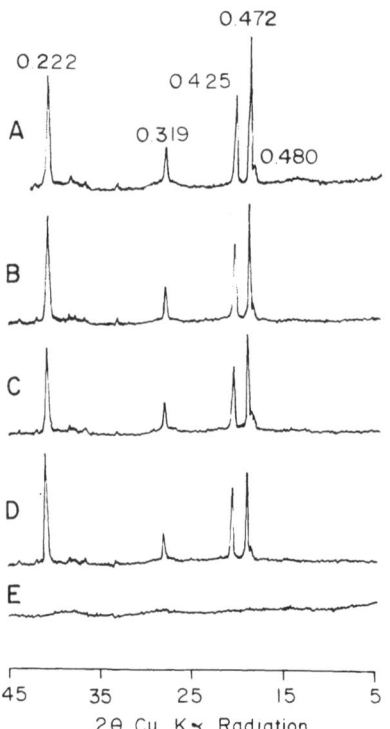

Figure 7. X-ray diffraction patterns of hydrolytic products of Al precipitated at the initial Al concentration of 1.1×10^{-3} M, OH/Al molar ratio of 3.0 and aged for 1 day at room temperature in the presence of (A) no organic acid, (B) 10^{-6} M p-hydroxybenzoic acid, (C) 10^{-6} M aspartic acid, (D) 10^{-6} M malic acid, and (E) 10^{-6} M citric acid (Kwong and Huang, 1979b; reprinted with permission of Elsevier Science Publishers B.V.). d-Spacings are in nanometers.

sufficiently to hamper the formation of bayerite after the 1-day aging (Figure 7B–D). When the concentration of citric, malic, and aspartic acids is raised to 10^{-4} M in the system, the hydrolytic products formed are noncrystalline to poorly crystalline to X-rays even after the 40-day aging (Figure 8B–D). p-Hydroxybenzoic acid, which formed the least stable complexes with Al (Table 7), does not inhibit the crystallization of hydrolytic precipitation products of Al (Figure 8A). The critical molar ratio of organic acid to Al in inhibiting the crystallization of Al hydroxides varies with the nature of organic acids (Kwong and Huang, 1979b). In the case of citric, malic, and aspartic acids, even at the molar ratio of organic acid to Al of 0.01, the crystallization of Al hydroxides is greatly retarded, and the presence of pseudoboehmite in the reaction products is indicated by the X-ray diffraction data.

Organic acids of higher molecular weight, such as tannic acid and fulvic acid, perturb the crystallization of precipitation products of Al. The noncrystalline to poorly crystalline hydrolytic precipitation products of Al, including pseudoboehmite, are formed in the presence of tannic acid (Kwong and Huang, 1981) and are shown to be fine, shapeless, hollow colloids that are deformed and aggregated (Figure 9). Kodama and Schnitzer (1980)

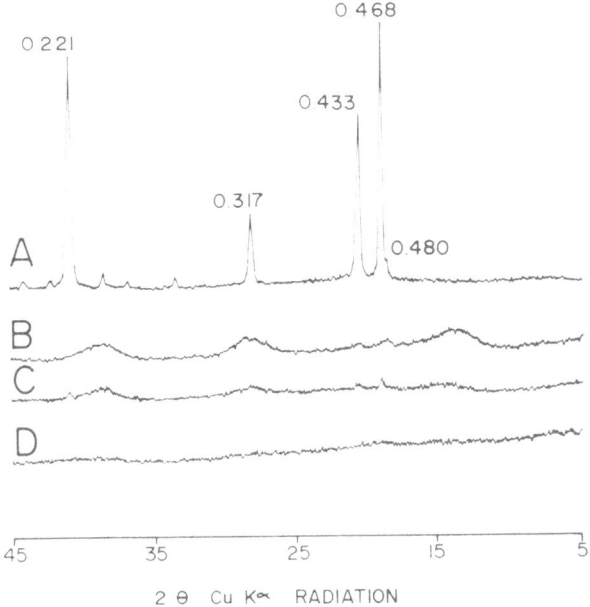

Figure 8. X-ray diffraction patterns of hydrolytic products of Al precipitated at the initial Al concentration of 1.10×10^{-3} M, OH/Al molar ratio of 3.0, and aged for 40 days at room temperature in the presence of 10^{-4} M: (A) p-hydroxybenzoic acid, (B) aspartic acid, (C) malic acid, and (D) citric acid (Kwong and Huang, 1979b; reprinted with permission of Elsevier Science Publishers B.V.). d-Spacings are in nanometers.

reported the significant effects of fulvic acid on the crystallization of Al hydroxides. In the absence of fulvic acid, gibbsite was formed at pH 6, a mixture of nordstrandite and bayerite was formed at pH 8, and bayerite crystallized at pH 10 (Table 8). At pH 6 and 8, the addition of increasing amounts of fulvic acid first delays and then inhibits the crystallization of these Al hydroxide polymorphs but favors the crystallization of pseudo-boehmite. As the ratio of fulvic acid to Al reaches 0.1, crystalline materials are no longer formed, and only amorphous precipitates are formed. At pH 10, the addition of fulvic acid totally inhibits precipitation and crystallization of hydrolytic products of Al.

Fulvic acid resembles low-molecular-weight aliphatic acids such as citric and malic acids in that it contains COOH and aliphatic OH groups. It also resembles quercetin and tannic acid, as it contains phenolic hydroxyl and ketonic C=O groups. The crystallization of Al hydroxides is delayed or inhibited by the presence of low-molecular-weight organic acids (Kwong and Huang, 1975, 1977, 1979b) and tannic acid (Kwong and Huang, 1981). The polymerization of Al hydroxides is inhibited by the presence of very

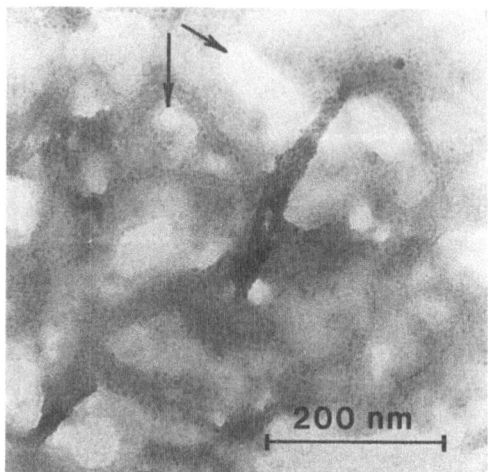

Figure 9. Transmission electron micrographs of the precipitation products of Al formed after the 40-day aging at room temperature at the initial concentration of 1.1×10^{-3} M Al, OH/Al molar ration of 3.0, and in the presence of 1.0×10^{-4} M tannic acid. Porous nature (see arrows) results from influence of tannic acid (Kwong and Huang, 1981; reprinted with permission of Elsevier Science Publishers B.V.).

small amounts of the flavone quercetin (Lind and Hem, 1975). Through these functional groups, fulvic acid can form stable complexes with Al and prevent the crystallization of Al(OH)$_3$ (Kodama and Schnitzer, 1980). The influence of fulvic acid on the crystallization of Al hydroxides also resembles that of salts (Hsu, 1967) in that fulvic acid appears to favor the formation of Al-O-Al (oxo) over that of

$$Al \underset{OH}{\overset{OH}{<}} Al$$

(01) linkages.

Pseudoboehmite formed at low temperatures and pressures in the absence of foreign anions is unstable and rapidly converted to Al(OH)$_3$ polymers (Souza Santos *et al.*, 1953; Aldcroft *et al.*, 1969; Vedder and Vermilyea, 1969; Yoldas, 1973; Serna *et al.*, 1977). However, certain organic and inorganic ligands not only promote the formation of pseudoboehmite over Al hydroxides, but also stabilize it (Hsu, 1967; Kwong and Huang, 1979b; Violante and Jackson, 1979, 1981; Kodama and Schnitzer, 1980; Violante and Violante, 1980; Kwong and Huang, 1981). Pseudoboehmite samples formed in the presence of various organic and inorganic perturbing ligands that are common in soil environments differ significantly in their

Table 8. Summary of X-ray data for the effect of FA on the crystallization of aluminum hydroxides at 30°C[a]

FA(mg)/500 ml[b] pH	0	0.25	0.5	1.0	1.5	2.5	5.0	50.0
6	GIBBS	gibbs + "bm"	gibbs + "bm"	gb + "boehm"	/	"BOEHM"	"BOEHM"	/
8	NORD + bayer	nord + bayer+ "bm"	nord + bayer+ "boehm"	/	ns and/or by (?) + "BOEHM"	/	ns and/or by (?) + "BOEHM"	Am. ppt.
10	BAYER	/[c]	—[d]	/	/	—	—	—

Source: Kodama and Schnitzer (1980); reprinted with permission of Elsevier Science Publishers.

[a] Abbreviations used in this table:

	Semi-quantitative estimates		
	Major	Minor	Trace
Bayerite	BAYER	bayer	by
Gibbsite	GIBBS	gibbs	gb
Nordstrandite	NORD	nord	ns
Pseudoboehmite	"BOEHM"	"boehm"	bm

Amorphous precipitate (Am. ppt.)

[b] Approximate FA/Al molar ratios ranged from 0 (0 mg FA) to 10^{-1} (50.0 mg FA).

[c] Not determined.

[d] No or very little precipitate.

Table 9. Comparison of X-ray diffraction data of standard boehmite and pseudoboehmite samples formed in the presence of optimum concentration range of perturbing organic and inorganic ligands at room temperature

Sample	d-Spacings (nm)						
	(020)	(120)	(140,031)	(051)	(200)	(231)	(251)
Boehmite (standard)[a]	0.611	0.316	0.235	0.186	0.185	0.145	0.131
Pseudoboehmite (citrate)	0.665	0.318	0.235	0.190	0.186	0.144	0.134
R[b] = 0.01	(β = 3.20)	(β = 2.35)	(β = 2.50)	b.v.[d]	(β = 2.00)	b.v.	b.v.
Pseudoboehmite (tartrate)	0.663	0.316	0.239	0.191	0.189	0.146	0.134
R = 0.01	(β = 3.00)	(β = 2.00)	(β = 2.40)	b.v.	(β = 1.80)	b.v.	b.v.
Pseudoboehmite (tannate)	0.707	0.319	0.235	0.190	0.185	0.152	0.136
R = 0.01	(β = 3.60)	(β = 2.40)	(β = 2.80)	b.v.	(β = 2.20)	(β = n.d.)	b.v.
Pseudoboehmite (tannate)	0.716	0.319	0.234	—	0.185	0.143	0.136
R = 0.02	(β = 3.65)	(β = 2.80)	(β = 3.00)	—	(β = 2.30)	(β = n.d.)	b.v.
Pseudoboehmite (chloride)	0.665	0.319	0.239	0.194	0.191	0.143	0.136
R = 700	(β = 3.55)	(β = 3.40)	(β = 2.80)	b.v.	(β = 2.30)	(β = 2.00)	b.v.
Pseudoboehmite (sulfate)	very broad	0.320	0.235	0.192	0.187	0.143	0.134
R = 70	(β = n.d.)[c]	(β = 4.00)	(β = 3.30)	(β = n.d.)	(β = n.d.)	(β = n.d.)	b.v.

Source: Violante and Huang (1984); reprinted with permission of the Soil Science Society of America.
[a] Only the d-spacings with a $I/I_1 \geq 15$ are reported (ASTM card no. 21-1307); β = breadth (2 θ) measured at half-maximum intensity; β and 2 θ position were measured according to the procedures described by Tettenhorst and Hofmann (1980).
[b] R is the initial ligand/Al molar ratio.
[c] Not determinable (the peak is visible, but the β value is difficult or impossible to determine).
[d] Barely visible (the peak is only barely visible).

Table 10. Order of effectiveness of organic ligands in promoting the formation of pseudoboehmite and X-ray noncrystalline precipitation products of Al after 5 months of aging at room temperature

Ligand	R[a]	R[b]
Phthalate ≅ succinate	7–14	15
Glutamate	0.4–3.0	4
Aspartate	0.1–2.0	2.5
Oxalate	n.c.p.[c]	1
Salicylate ≅ malate	0.02–0.05	0.10
Tannate	0.01–0.03	0.04
Citrate	0.007–0.02	0.03
Tartrate	0.005–0.015	0.02

Source: Violante and Huang (1985); reprinted with permission of Clay Minerals Society.

[a] R = the range of the initial ligand/Al molar ratios at which stable pseudoboehmite was formed at pH 8.2.

[b] R = the lowest initial ligand/Al molar ratio at which X-ray noncrystalline materials were formed at pH 8.2.

[c] No evidence for the crystallization of pure pseudoboehmite.

surface morphology and chemical composition (Violante and Huang, 1984). The pseudoboehmite samples that are synthesized at room temperature in the presence of selected organic and inorganic ligands show similar X-ray diffraction data (Table 9). The Al oxyhydroxides consist of fibrillar particles as well as many aggregates (> 5 nm) even after ultrasonification. These studies throw new light on the genesis of noncrystalline alumina of boehmite characteristics in certain tropical soils (De Villiers, 1969) and of some bauxites that consist of submicroscopic boehmitelike particles that are very similar to pseudoboehmite (Lippens and Steggerda, 1970).

The sequence of the relative effectiveness of certain perturbing organic ligands commonly present in nature in promoting the formation of pseudoboehmite over $Al(OH)_3$ polymorphs is as follows: phthalate ≅ succinate < glutamate < aspartate < salicylate ≅ malate < tannate < citrate < tartrate (Table 10). As the molar ratios of perturbing ligands to Al exceed the critical ratios for the formation of pseudoboehmite, organic ligands that are coprecipitated with Al inhibit the crystallization of Al hydroxides or oxyhydroxides and promote the formation of X-ray noncrystalline Al precipitation products. First of all, the hydrolysis of the positively charged sites of hydroxy-Al polymers is retarded because of the occupation of the coordination sites of Al by the perturbing ligands instead of H_2O molecules (Kwong and Huang, 1979a). Secondly, because of steric factors, the perturbing ligands occupying the coordination sites of Al distort the arrangement of the platelets (the *001* sheets) normally found in crystalline Al hydroxides (Figure 10).

The order of common organic ligands in promoting the formation of

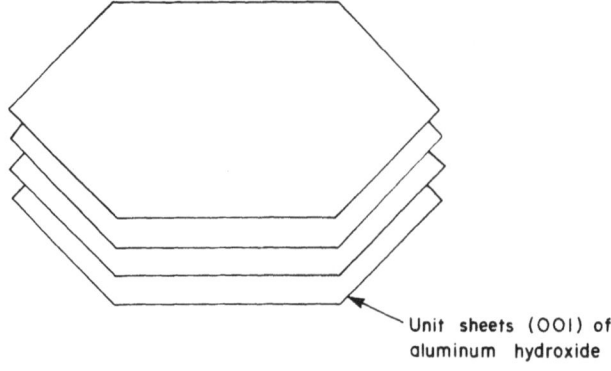

Unit sheets (OOl) of
aluminum hydroxide

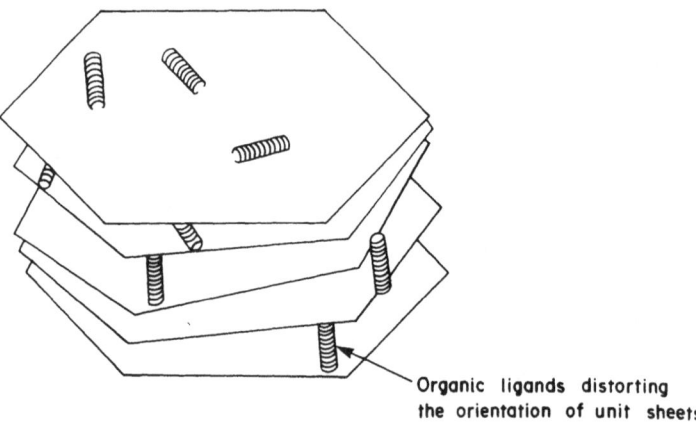

Organic ligands distorting
the orientation of unit sheets

Figure 10. Structural models showing the distortion of the orientation of the
platelets (the *001* sheets) of Al hydroxides by organic ligands which coprecipitate
with Al and are incorporated into the sheets (modified from Kwong and Huang,
1979b; reprinted with permission of Elsevier Science Publishers B.V.).

noncrystalline Al precipitation products is as follows: phthalate $\cong$ succinate $<$ glutamate $<$ aspartate $<$ oxalate $<$ salicylate $\cong$ malate $<$ tannate $<$
citrate $<$ tartrate (Table 10). Many organic ligands such as tartrate, citrate,
tannate, malate, and salicylate are much stronger than inorganic ligands
such as silicate and phosphate in promoting the formation of precipitation
products of Al that are noncrystaline to X-rays (Violante and Huang,
1985). The strength of the ligands promoting the formation of noncrystalline materials appears to vary with their nature, size, and affinity for Al.
Ligands that form polydentate complexes generally have a stronger ability
in causing the structural distortion of precipitation products of Al than
those with fewer functional groups. The implications of these findings are

evident in the formation of short-range ordered organomineral complexes in soils (Mitchell *et al.*, 1964; Campbell *et al.*, 1977). Furthermore, these studies indicate that besides hydroxy-Al interlayering of expansible phyllosilicates (Jackson, 1963b; Rich, 1968; Bertsch and Barnhisel, 1987), the formation of crystalline Al hydroxides in soils and sediments can be hampered through the complexation reactions of hydroxy Al with these organic acids, especially in environments where accumulation of organic acids tends to occur. Substantial concentrations of crystalline Al hydroxides are thus absent in organic matter–rich soils in the temperate regions.

D. Formation of Aluminum Hydroxide Polymorphs as Influenced by Organic Ligands

The nature and concentration of organic acids and the associated pH are related to the formation of Al hydroxide polymorphs (Table 11). At a given pH and organic acid/Al ratio, there is a change in the final aluminous products formed from bayerite to nordstrandite and/or gibbsite and finally to pseudoboehmite and noncrystalline materials, approximately according to the increasing chelating power of the organic acids. The same order of formation of the various aluminous precipitation products follows at a given pH by increasing the concentration of some of the organic acids.

Succinic acid weakly complexes Al ions (Murmann, 1964) by forming unstable seven-membered rings, so its presence has a relatively minor influence on the crystallization rate, particularly at pH > 8.0 (Table 11). Data for phthalic acid show the same trend as for succinic acid (not reported in Table 11).

Glycine (aminoethanoic acid), a bidentate agent, like phthalic and succinic acids, complexes Al ions by forming a more stable five-membered ring. Consequently, in glycine systems, particularly at the highest glycine/Al ratio and at pH 8.0 and 9.0, the rate of Al $(OH)_3$ crystallization is slower than in succinate or phthalate systems (Violante and Violante, 1980), and the formation of gibbsite (at pH 8) and nordstrandite (at pH 9.0) is favored (Table 11).

With more effective retarding agents, such as glutamic and aspartic acids, gibbsite, nordstrandite, and pseudoboehmite form easily at pH 8.0 and commonly at pH 9.0 (Table 11). Aspartic and glutamic acids are both tridentate, but in the presence of the former, the crystallization rates of hydrolytic products of aluminum are slower than in the presence of the latter. In samples where aspartic acid is present, gibbsite, nordstrandite, and pseudoboehmite form more easily than bayerite. Aspartic acid, in fact, acts as a stronger retarding agent, probably because the COOH group at the end of the aliphatic side chain and the $-NH_2$ group might stabilize the complex by chelating Al ions with two stable five- and six-membered rings. On the contrary, glutamic acid might complex Al ions more weakly by forming a stable five-membered ring and an unstable seven-membered

Table 11. Aluminum hydroxides and oxyhydroxides formed in the presence of complexing organic acids after 60 days at 20°C

Organic acid/Al molar ratio	Acid								
	Succinic	Glycine	Glutamic	Aspartic	Oxalic	Malic	Salicylic	Citric	Tartaric
Samples aged at pH 8.0									
0.014	B, (G)	B, (G)	—	G	G, (P)	G, P	—	P	P
0.029	B, (G)	B, (G)	G, N, (B)	P, G, (N)	G, P or P	P	P	A	A
0.050	B, (G)	B, G	G, N	P, (G)	P	P	P	A	A
0.167	G, N, B	G	P	A	A	A	A	A	A
Samples aged at pH 9.0									
0.029	B	B, (N)	B, N	N	N, B	N, P	N, P	P	P
0.050	B	B, N	N	G, N, P	N, (P, B)	P, (N, G)	P, N, (G)	P	A
0.167	B, N	N, (P)	G, N, P	G, P	G, P or P	P	P, (N, G)	A	A
Samples aged at pH 10.0									
0.050	B	B	B	B	B, (P)	B, N	N, (B)	N	P
0.167	B	B	B, (P)	B, P	N, B, (P)	N, P	N, P, (G)	P	A

Source: Violante and Violante (1980); reprinted with permission of Clay Minerals Society.

A = amorphous Al hydroxides; P = pseudoboehmite; B = bayerite; G = gibbsite; N = nordstrandite; ( ) small amounts; — = no Al hydroxides or oxyhydroxides detected.

ring. According to Das Sarma (1956), glutamate, aspartate, and other α-amino acid ligands do not always easily form a tridentate ligand to a single ion, and other structures have been suggested as alternatives where these ligands behave as either tridentate or bidentate ligands. However, the above reasoning easily explains the greater influence of aspartic acid over glutamic acid on the crystallization of $Al(OH)_3$ polymorphs (Table 11).

Gibbsite is commonly favored over nordstrandite when the crystallization rate is particularly slow, even at high pH values. At pH 9.0 and at a glutamic acid/Al ratio of 0.1, almost pure nordstrandite has been synthesized, whereas in the presence of aspartate, gibbsite and nordstrandite are formed (Violante and Violante, 1980). In addition, a concentration increase of a complexing organic ligand results in a progressive increase in the ratio of gibbsite to nordstrandite (Table 11).

In the optimal concentration range, oxalic acid has a greater influence than malonic, succinic, and glutaric acids (in the order listed) in favoring the formation of gibbsite, nordstrandite and/or pseudoboehmite (Violante and Violante, 1980). This may be explained by considering that chelates of five- or six-membered rings are more stable than larger ones. The closer the carboxyl groups, the more active is the bicarboxylate anion. The carboxylate group in glutarate, and probably in higher homologues, are too far apart; hence these ligands behave like monocarboxylates—e.g., acetate (Violante and Violante, 1978).

Strong chelating organic acids significantly retard the crystallization process, so that gibbsite is favored as the precipitation product even at high pH (Violante and Violante, 1980). Intermediate conditions created by moderately strong chelating anions do not retard the rate of crystallization significantly to favor gibbsite but rather favor nordstrandite, particularly at pH > 8.0. In the absence of organic acids or in the presence of weakly complexing organic acids, the rapid crystallization processes lead to the formation of bayerite. The data indicate that organic acids affect the kinetics of the crystallization of hydrolytic products of Al and thus influence the kinds of Al hydroxide polymorphs formed. In alkaline soils, the presence of clays and organic matter with its carboxylic and amino groups would inhibit the crystallization of bayerite by favoring the formation of gibbsite or nordstrandite. These results provide the interpretation that nordstrandite and gibbsite, rather than bayerite, are common in some alkaline soils. Furthermore, it is logical that bayerite is virtually absent in soils and natural environments (Schoen and Roberson, 1970; Milton *et al.*, 1975).

E. Formation of Hydroxy-Aluminum-Organics-Clay Complexes

Soils are a complex mixture of organic and inorganic components that interact with each other (Greenland, 1965a,b; Schnitzer and Kodama, 1977). The stability of the chloritelike complexes in montmorillonite in the alkaline range of pH is influenced by the presence of citric acid (Violante and Violante, 1978; Violante and Jackson, 1979, 1981).

Table 12. Total Al sorbed from solution, KCl-exchangeable Al, HCl-extractable Al, and CEC of montmorillonite after equilibration with hydroxy Al and citric acid

Sample No.		NaOH/Al	Citrate/Al	pH Initial	pH Final	Al[a] sorbed	Sorbed Al[b] exch. by 1 M KCl	Sorbed Al[c] ext. by 0.2 M HCl	CEC[d] before extraction by 0.2 M HCl	CEC[d] after extraction by 0.2 M HCl
							g · kg^{-1}		cmol (1/2 Ca^{2+}) · kg^{-1}	
1	In water			7.4	7.4	ND[e]	ND	ND	96	95
2	In AlCl$_3$	0[f]		4.0	3.9	7.54	5.48	7.35	78	83
3		0	0.1	3.5	3.6	6.35	5.71	5.95	92	88
4		0	0.5	3.0	3.0	6.07	6.06	6.31	84	86
5		0	1.0	2.8	3.0	5.65	5.34	5.77	89	85
6		2.5	0	4.5	4.1	49.95	1.22	10.49	31	43
7		2.5	0.1	4.6	4.7	48.43	0.82	1.85	34	47
8		2.5	0.5	4.3	4.6	28.18	5.55	8.35	61	76
9		2.5	1.0	3.6	3.7	14.96	6.33	6.76	82	83

Source: Goh and Haung (1984); reprinted with permission of the Agricultural Institute of Canada.

[a] Mean error of duplicates = ± 0.43 g Al · kg^{-1}.
[b] Mean error of duplicates = ± 0.21 g Al · kg^{-1}.
[c] Mean error of duplicates = ± 0.70 g Al · kg^{-1}.
[d] Mean error of duplicates = ± 3 cmol (1/2 Ca^{2+}) · kg^{-1}.
[e] ND, not detectable.
[f] At NaOH/Al = 0, no NaOH was added to the montmorillonite suspension in the AlCl$_3$ solution.

More recent research data show that the formation of hydroxy-Al-montmorillonite complexes in the acidic pH range that is most conducive to chloritization (Rich, 1968) is significantly perturbed by low-molecular-weight organic acids such as citric and tannic acids (Goh and Huang, 1984, 1986). Once hydroxy Al ions interact with organic acids to form hydroxy Al-organic acid complexes, the adsorption of these complexes by the inter-layer spaces of montmorillonite can take place only if the net attractive forces between the positively charged sites of the complexes and negatively charged clay surfaces can overcome the repulsive forces between the anionic ligands of the complexes and clay surfaces and any steric hindrances that may be present. Table 12 shows that the citrate/Al molar ratio has an important influence on the formation of hydroxy Al interlayers. At the citrate/Al molar ratio of 0.1 or less, the mutual repulsion between negatively charged sites is apparently not strong enough to prevent the formation of hydroxy Al interlayers in smectite. Between the citric acid/Al molar ratios of 0.3–0.5, a smaller but substantial amount of Al can still be adsorbed by smectite. If the citric acid/Al molar ratio is 0.7 or higher, very little is adsorbed by smectite. The formation of hydroxy-Al-smectite complexes appears to be critically perturbed if the citrate/Al molar ratio exceeds 0.5.

The polynuclear Al ion of double ring structure (Hsu, 1977) can be chelated by citrate (Kwong and Huang, 1979a), and its structure can be depicted below:

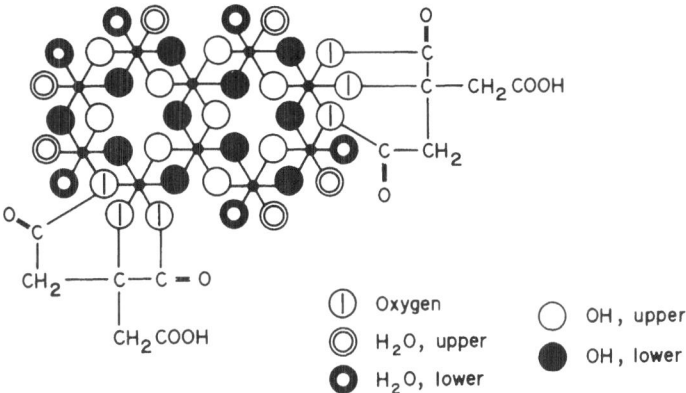

In view of the structure of a hydroxy-Al-citrate as illustrated above and the data obtained by using cation and anion exchange resins (Goh and Huang, 1986), hydroxy-Al-citrate chelates can be sorbed on the interlayer spaces of smectite. The adsorption of hydroxy-Al-citrate by smectite causes structural distortion from within the interlayer spaces of the clay as indicated by the broadening of the (001) peak in X-ray diffractograms (Figure

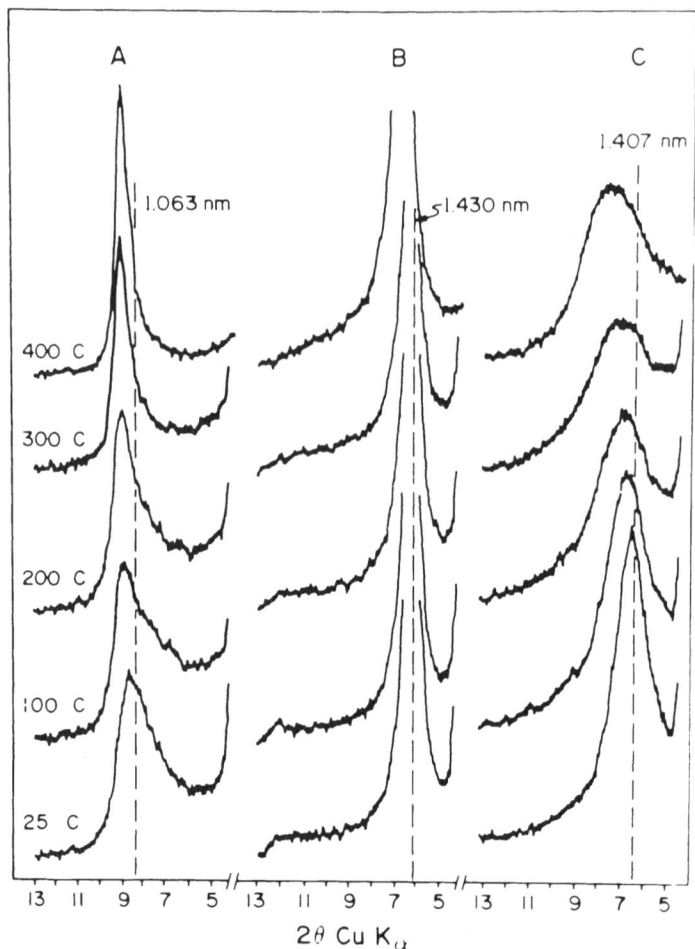

Figure 11. The X-ray diffractograms of the first-order reflection of selected solid-phase reaction products (room temperature) after heating at various temperatures: (A) Na-montmorillonite; (B) hydroxy-aluminum-montmorillonite, NaOH/Al molar ratio of 2.5; and (C) hydroxy-aluminum-citrate-montmorillonite, NaOH/Al molar ratio of 2.5, citrate/Al molar ratio of 0.5 (Goh and Huang, 1984; reprinted with permission of the Agricultural Institute of Canada).

11C) and the obscuring of the IR absorption bands at 3695 and $3570 cm^{-1}$ (Figure 12C) representing the presence of hydroxy Al interlayers. In addition, the dehydroxylation endotherm is spread over a wide range of temperatures ranging from 430°C to about 520°C as the citrate/Al ratio is increased from 0.1 to 0.5 (Goh and Huang, 1985). Obviously, the perturbation to the interlayer hydroxy-Al caused by citric acid has the effect of causing the interlayers to lose their structural water more slowly. The com-

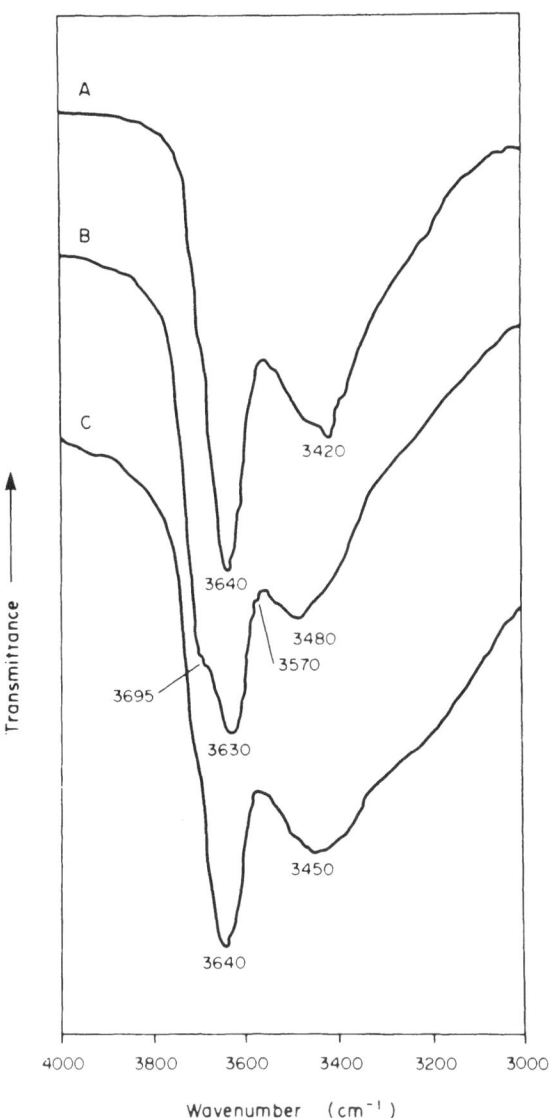

Figure 12. Infrared spectra of selected solid-phase reaction products (room temperature): (A) Na-montmorillonite aged 3 months in water; (B) hydroxy-aluminum-montmorillonite aged at initial pH of 5.0; and (C) hydroxy-aluminum-citrate-montmorillonite aged at initial pH of 5.0 and citrate/Al molar ratio of 0.5 (Goh and Huang, 1984; reprinted with permission of the Agricultural Institute of Canada).

plexation between tannic acid and Al is deemed to occur more favorably when the pH of the suspension is increased, since the reaction generates protons (Goh and Huang, 1986). The protons are not likely generated by the simple dissociation of phenolic hydroxyls as the initial pH of aging suspension ranges from 5 to 6 and thus well below the reported pKa of phenolic groups, which is about 9.0 (Weast, 1978). The formation of a five-membered ring in the complex is more stable than a monodentate complex that has Al being coordinated to the carboxylate, COO^- group (Cotton and Wilkinson, 1980). Further growth of the hydroxy-Al-tannate complex to a large hydroxy-Al-tannate complex polymer may be possible (Goh and Huang, 1986). The high molecular weight of tannic acid apparently causes the hydroxy-Al-tannate complexes to precipitate as a discrete phase which acts as cementing agents between clay particles. The forces of physical adsorption are additive in high-molecular-weight organic compounds (Greenland, 1965a; Kwong and Huang, 1981). The aggregation of hydroxy-Al-clay particles is thus promoted by tannic acid.

The finding on the perturbation of Al interlayering of expansible phyllosilicates by organic acids also provides an interpretation to the observation that the accumulation of hydroxy Al interlayers is more pronounced under well-drained conditions than in imperfectly drained soils (Jackson, 1963b; Rich, 1968; Huang and Lee, 1969), since the latter environment is more conducive to the accumulation of organic acids (Stevenson, 1967).

More recent studies show that the formation of hydroxy-Al-montmorillonite complexes is significantly perturbed by HAs (Singer and Huang, 1986). The observation of Rich (1968) that Al interlayering of clays is not favorable in soils that are high in organic matter is thus explained.

F. Pedogenic Significance of Organic Ligands in the Formation of Short-Range Ordered Aluminosilicates

Frequency distribution of allophanes and imogolite in soils and sediments has been reported (Wada, 1977, 1980; Tait et al., 1978; Ross and Kodama, 1979; McKeague and Kodama, 1981; Farmer et al., 1980; Anderson et al., 1982). These short-range ordered aluminosilicates have large specific surface and high chemical reactivity (Wada, 1981). They have a significant effect on the physical, chemical, and biological properties of soils and sediments. Little is known of the effect of low-molecular-weight organic acids on the formation of these short-range ordered aluminosilicates (Farmer, 1981). Very recently, Inoue and Huang (1984a, 1985) reported that citric acid (molar ratio of citric acid to Al at ≤ 0.1) greatly perturbs the interactions of hydroxy Al ions with orthosilicic acid and thus hinders the formation of imogolite and allophanes.

In the presence of citric acid and at a Si/Al ratio of 0.5 and an OH/Al ratio of 1.0–2.0, protoimogolite is found in the soluble products that pass through the 0.01-μm pore-size membrane filter and are recovered by freeze-drying the filtrates (Inoue and Huang, 1985). Protoimogolite has positive charges in the acidic condition (Farmer, 1981); therefore, protoimogolite, which is composed of < 10-nm size particles (Inoue and Huang, 1984a), apparently forms complexes with citric acid. Thus, the complexed citric acid impedes the conversion of protoimogolite to imogolite.

The SiO_2/Al_2O_3 molar ratio of the precipitates formed is greatly influenced by citric acid and decreases markedly with increasing the citric acid/Al molar ratio (Inoue and Huang, 1985). Inverseley, the organic C content, which is a measure of the amount of citrate in the precipitate, increases with an increase in the citric acid/Al molar ratio. The results clearly show that citric acid coprecipitates with aluminosilicates during their formation.

The XRD pattern of the precipitates formed in the absence of citric acid indicates that imogolite (d = 2.16, 1.51, 0.87, and 0.64 nm) (Farmer and Fraser, 1979; Wada et al., 1979) and small amounts of boehmite (d = 0.61 nm) and bayerite (d = 0.47 nm) (Hsu, 1977) are dominant in the products (Figure 13). The intensity of the XRD peaks of imogolite, however, is significantly reduced with increasing citric acid/Al ratios.

The IR spectra of the precipitates formed in the absence of citric acid (Figures 14a,d, 15a) resemble those obtained from natural imogolite. The characteristic absorption bands of imogolite at 995, 936, 700, 565, 504, 422, and 345 cm^{-1} (Wada and Harward, 1974; Wada, 1977, 1980; Farmer et al., 1977, 1979; Farmer and Fraser, 1979) are distorted or weakened with increasing citric acid/Al ratios of the parent solution (Figures 14a–c, 14d–h, 15a–c).

The IR spectra of the precipitate formed at the Si/Al molar ratio of 1.0, OH/Al molar ratio of 1.0 and citric acid/Al molar ratio of 0.02 (Figure 14c) show absorption bands at 965, 563, 424, and 342 cm^{-1}, which are very similar to those of protoimogolite (Farmer et al., 1978, 1979; Farmer and Fraser, 1979; Inoue and Huang, 1984a). Protoimogolite is, however, known to be water-soluble. Therefore, these precipitates may be termed ill-defined aluminosilicate complexes.

The IR spectra of the precipitates formed in the Si/Al molar ratio of 1.0 and the OH/Al molar ratio of 3.0 show a broad absorption maximum at 990 cm^{-1} (Figure 15d), which is a feature common to allophanic clay separated from Ando soils (Wada and Harward, 1974; Wada, 1977, 1980). With increasing citric acid/Al ratio of the parent solution from 0.01 to 0.03, however, the IR spectra of the precipitates show a gradual shift of the Si-O stretching maximum from 985 to 965 cm^{-1} (Figure 15e–h). Especially at citric acid/Al ratios of 0.1–0.3 (Figure 15g,h), highly disordered products are formed, which are obviously different from allophane precipitated from the solution in the absence of citric acid. The disordered products show a sharp absorption maximum at 965 cm^{-1} and other absorption bands at 576, 424, and 345 cm^{-1} (Figure 15g,h), which are similar to those of

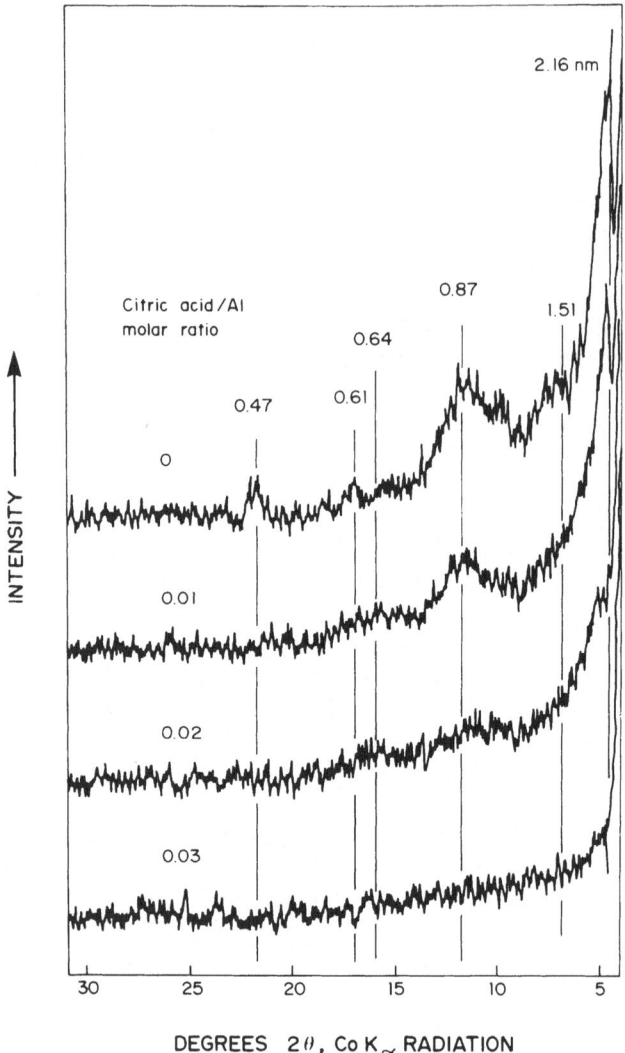

Figure 13. X-ray power diffractograms of the precipitates formed from solutions containing hydroxy Al ions, orthosilicic acid, and citric acid at Si/Al molar ratio of 0.5, OH/Al molar ratio of 2.0, and citric acid/Al molar ratios of 0–0.03 (Inoue and Huang, 1985; reprinted with permission of Clay Minerals Society).

protoimogolite allophane (Farmer *et al.*, 1980; Parfitt and Henmi, 1980). "Protoimogolite" allophane samples isolated from soils and pumices in New Zealand show IR spectra similar to that of protoimogolite, but their unit particles appear to be hollow spherules or polyhedra 3.5–5.0 nm in diameter with the SiO_2/Al_2O_3 ratio close to 1.0 (Parfitt and Henmi, 1980).

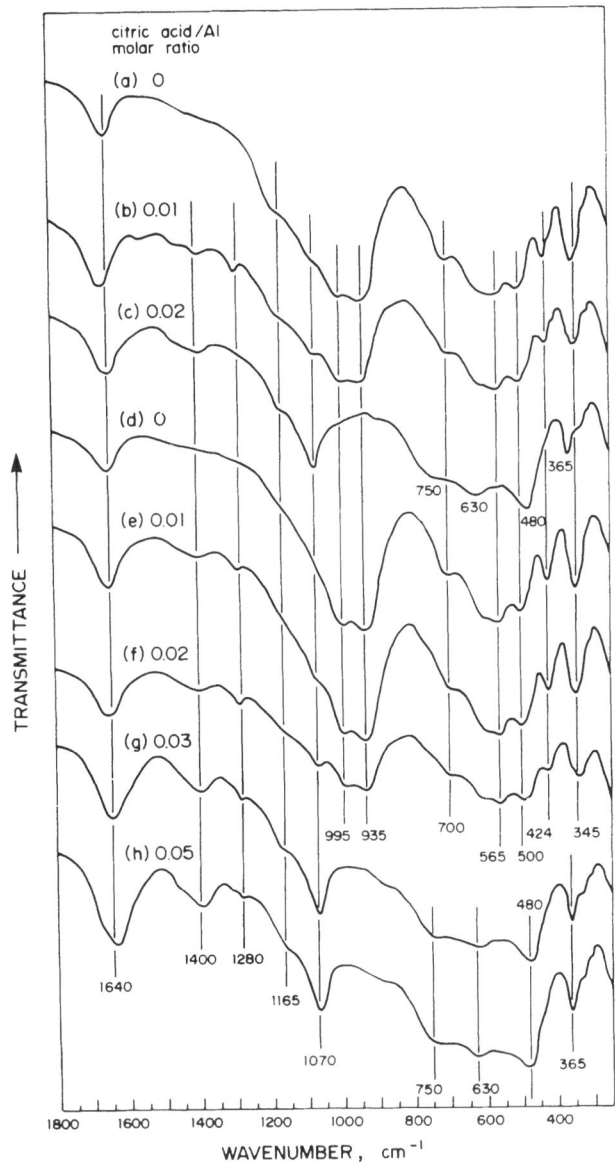

Figure 14. IR spectra of precipitates formed from solutions containing hydroxy Al ions, orthosilicic acid, and citric acid at citric acid/Al molar ratios of 0–0.05. (a–c) Si/Al molar ratio of 1.0 and OH/Al molar ratio of 1.0. (d–h) Si/Al molar ratio of 0.5 and OH/Al molar ratio of 2.0 (Inoue and Huang, 1985; reprinted with permission of Clay Minerals Society).

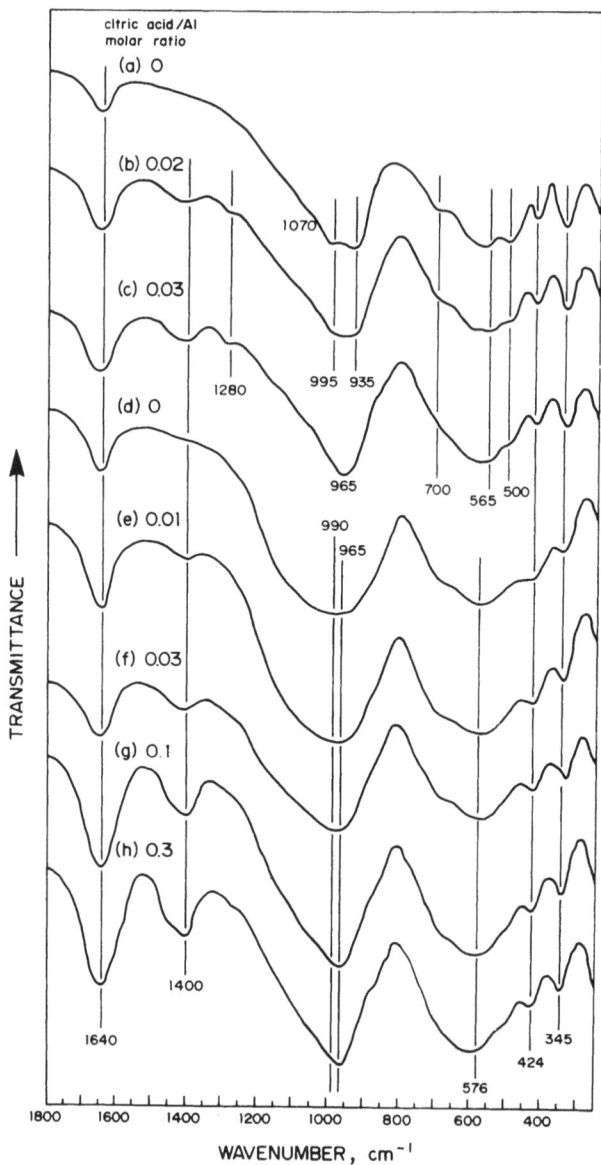

Figure 15. IR spectra of precipitates formed by reaction of hydroxy Al ions with orthosilicic acid both in the absence and presence of citric acid. (a–c) Si/Al molar ratio of 0.5 and OH/Al molar ratio of 2.8. (d–h) Si/Al molar ratio of 1.0 and OH/Al molar ratio of 3.0 (Inoue and Huang, 1985; reprinted with permission of Clay Minerals Society).

The precipitates formed by the reaction of hydroxy Al with orthosilicic acid in the presence of citric acid (Si/Al molar ratio of 1.0, OH/Al molar ratio of 2.8, citric acid/Al molar ratio of 0.1 or 0.3) are also very similar to protoimogolite in their IR spectra (Figure 15g,h). They are Al complexes of mixed ligands containing hydroxyl, citrate, and orthosilicate. The decrease of the SiO_2/Al_2O_3 ratio of the precipitates (Inoue and Huang, 1985) and the gradual shift of IR absorption maximum from 990 to 965 cm^{-1} (Figure 15g,h) seem to increase with the increase in the degree of complexation of hydroxyaluminosilicates with citrate ligands. This type of precipitates may be tentatively classified as ill-defined aluminosilicate complexes. Such groups of aluminosilicates seem to resemble the so-called allophanelike material (Wada and Greenland, 1970; Wada and Harward, 1974) and the ill-defined fraction of the allophane-imogolite complex (Farmer et al., 1983).

In the absence of citric acid, smooth and curved imogolite threads appear to be of micrometer length with the diameter of 2.0–50.0 nm (Figure 16a). A high-magnification TEM of the precipitates (Figure 16d) shows that the threads consist of fine tube units with inner and external diameters of about 1.0 and 2.0 nm, respectively. Irregular-shaped particles with different morphological characteristics and their aggregates are, however, observed in the precipitates formed at the Si/Al molar ratio of 0.5, OH/Al molar ratio of 2.0, and citric acid/Al molar ratio of 0.03 (Figure 16b), which are characterized by the predominance of disordered products with low SiO_2/Al_2O_3 ratios and pseudoboehmite.

A TEM of the precipitates formed at the OH/Al molar ratio of 2.8, Si/Al molar ratio of 0.5, and citric acid/Al molar ratio of 0.03 shows the presence of a gellike material (Figure 16c). The precipitates formed at the OH/Al molar ratio of 2.8, Si/Al molar ratio of 1.0, and citric acid/Al molar ratio of 0.02 are composed of gellike materials, very distorted imogolite tubes, and hollow spherules (Figure 16e).

Figure 17a shows a high-magnification TEM of noncrystalline precipitates formed at the OH/Al molar ratio of 3.0 and Si/Al molar ratio of 1.0 in the absence of citric acid. The presence of hollow spheres with diameters of 3.5–5.0 nm in the TEM (Figure 17a) and IR spectra of the products (Figure 15d) reveal that the noncrystalline aggregates observed in the TEM mainly consist of allophane. High-magnification TEM has shown that natural allophane in soils (Wada and Harward, 1974; Wada, 1977) and river sediments (Wells et al., 1977; Inoue et al., 1980) and synthetic allophane (Wada et al., 1979) consist of hollow spheres 3.5–5.0 nm in external diameter with walls 0.7–1.0 nm thick (Wada, 1979). The morphology of the hollow spheres is, however, markedly distorted by the presence of citric acid during their formation; only a few hollow spherules with diameters of 3.5–5.0 nm are present in the irregular aggregates (Figure 17b,c).

Organic ligands of different chelating power have varying effects on the formation of allophane and imogolite and the nature of precipitation

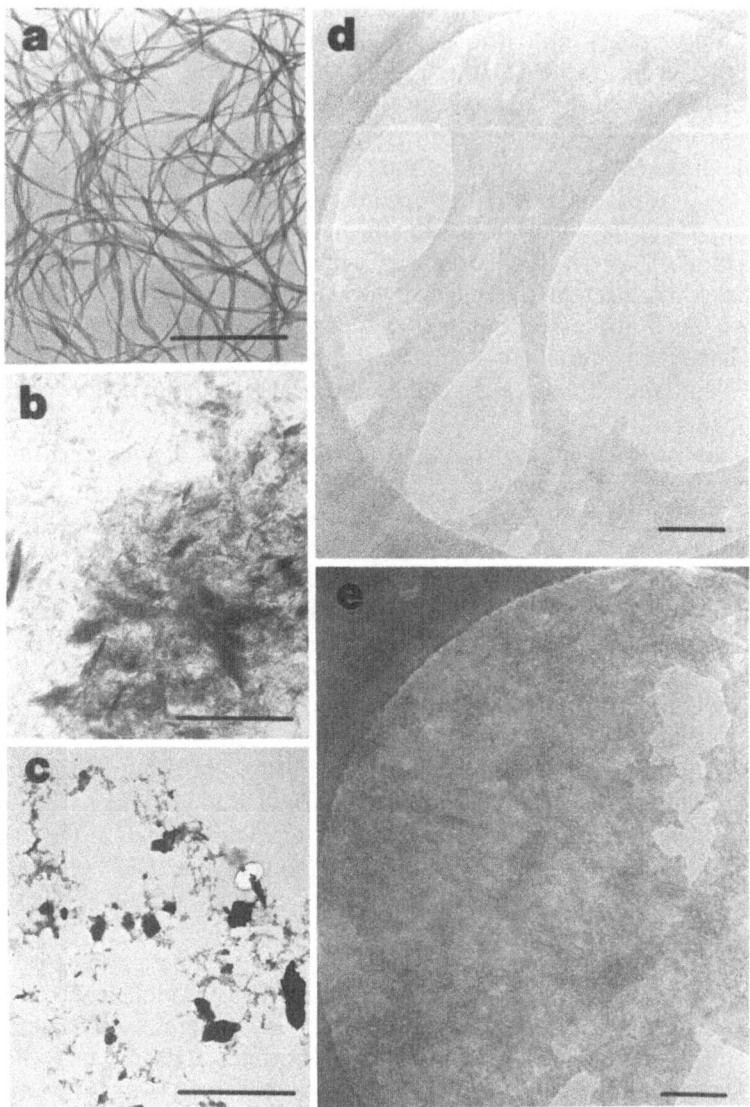

Figure 16. Transmission electron micrographs of precipitates formed by reaction of hydroxy Al ions with orthosilicic acid at citric acid/Al molar ratios of 0–0.03. (a) OH/Al molar ratio of 2.0, Si/Al molar ratio of 0.5, citric acid/Al molar ratio of 0. (b) OH/Al molar ratio of 2.0, Si/Al molar ratio of 0.5, citric acid/Al molar ratio of 0.03. (c) OH/Al molar ratio of 2.8, Si/Al molar ratio of 0.5, citric acid/Al molar ratio of 0.03. (d) OH/Al molar ratio of 2.0, Si/Al molar ratio of 1.0, citric acid/Al molar ratio of 0 (high magnification). (e) OH/Al molar ratio of 2.8, Si/Al molar ratio of 1.0, and citric acid/Al molar ratio of 0.02 (high magnification). Scale bar = 1μm (a–c), 50 nm (d,e) (Inoue and Huang, 1985; reprinted with permission of Clay Minerals Society).

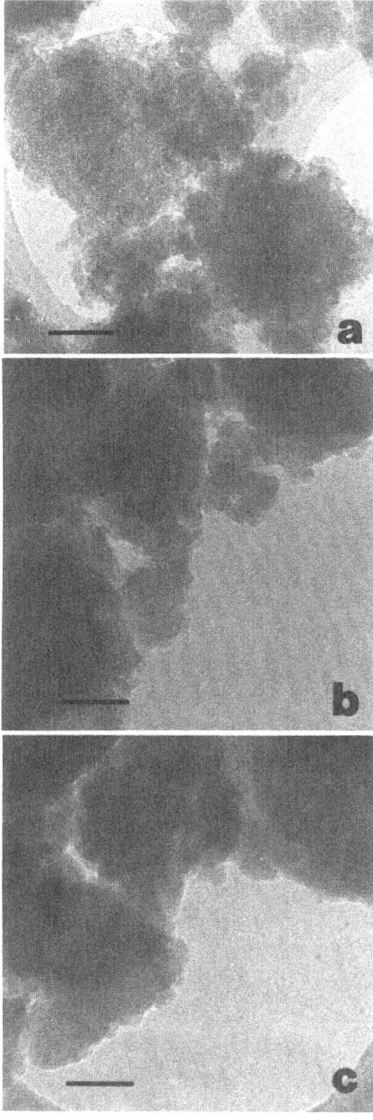

Figure 17. Transmission electron micrographs of precipitates formed by the reaction of hydroxy Al ions with orthosilicic acid at Si/Al molar ratio of 1.0 and OH/Al molar ratio of 3.0 in both the absence and presence of citric acid. (a) OH/Al molar ratio of 3.0, Si/Al molar ratio of 1.0, citric acid/Al molar ratio of 0. (b) OH/Al molar ratio of 3.0, Si/Al molar ratio of 1.0, citric acid/Al molar ratio of 0.1. (c) OH/Al molar ratio of 3.0, Si/Al molar ratio of 1.0, citric acid/Al molar ratio of 0.3. Scale bar = 50 nm (Inoue and Huang, 1985; reprinted with permission of Clay Minerals Society).

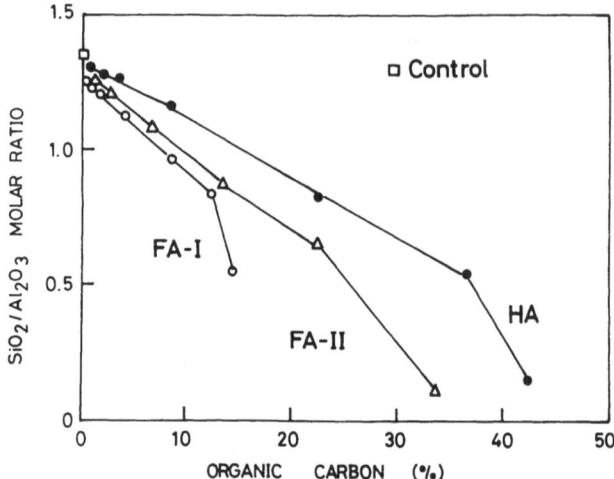

Figure 18. Relationship between SiO_2/Al_2O_3 molar ratios and organic carbon content of precipitates formed from solutions containing hydroxy Al ions, orthosilicic acid, and humic substances (Inoue and Huang, 1987; reprinted with permission of Clay Minerals Society).

(> 0.01 μm) and soluble (< 0.01 μm) products (Inoue and Huang, 1986). Chemical composition, size, number, and nature of the functional groups and concentration of organic ligands have a bearing on the perturbation of the formation of short-range ordered aluminosilicates. The order of effectiveness of organic ligands studied is tannic $\gg$ DL-tartaric $>$ citric $>$ DL-malic $\gg$ salicylic $>$ DL-aspartic $\geq$ p-hydroxybenzoic. Organic acids with a strong affinity for Al greatly perturb the interaction of hydroxy-Al ions with orthosilicic acid and thus retard or even inhibit the subsequent formation of allophane and imogolite.

More recent studies (Inoue and Huang, 1984b, 1987) show that FAs and HAs can also greatly perturb the genesis of allophane and imogolite. The SiO_2/Al_2O_3 ratio of the precipitates decreases with increasing organic C carbon (Figure 18). The strong competition of such humic substances with orthosilicic acid for the coordination sites of Al is considered to account for the marked impedance of the humic substances on the formation of allophane. The IR spectrum of the precipitates formed in the absence of humic substances (Figure 19a) gave strong absorption maxima at 1633, 988, and 570 cm^{-1} and weak absorption bands at 424 and 345 cm^{-1}. These bands are common to soil allophane (Wada and Harward, 1974; Wada, 1977, 1980). At increasing concentrations of humic substances, however, the absorption maxima of the Si-O stretching band at 900–1000 cm^{-1} shifted gradually from 988 to 944 cm^{-1} (Figure 19b–d) and from 983 to 964 cm^{-1} (Figure 19f–i), and their intensities all weakened (Figure 19). The IR bands (972–944, 569–554, 424, and 345 cm^{-1}) of the precipitates formed in

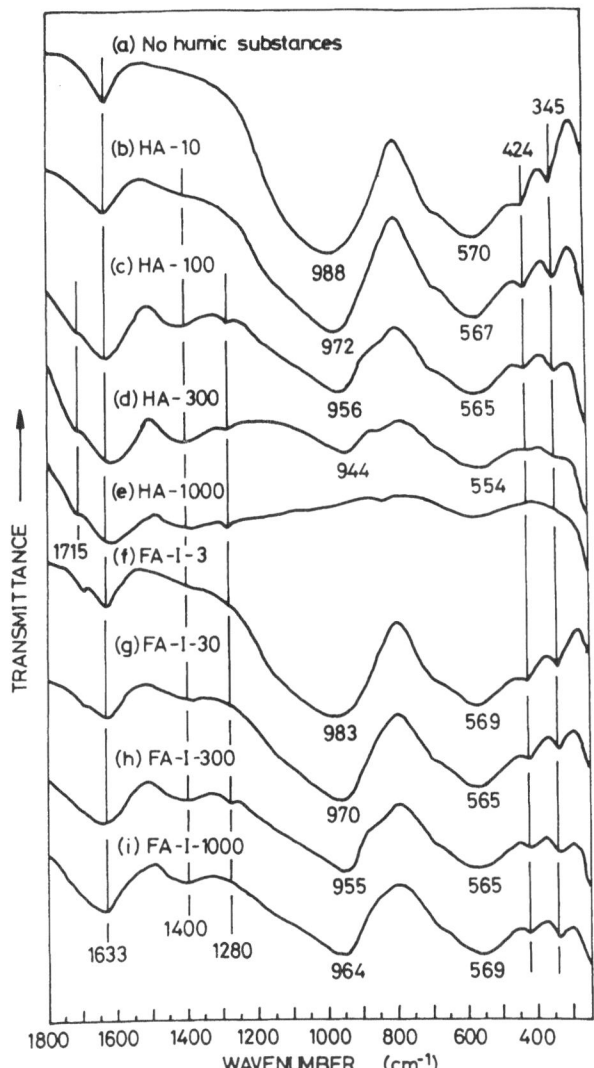

Figure 19. Infrared spectra of precipitation products (> 0.01 μm) formed by reaction of hydroxy Al ions with orthosilicic acid in both the absence and presence of humic substances: (a) no humic substances; (b) humic acid, 10 mg/L; (c) humic acid, 100 mg/L; (d) humic acid, 300 mg/L, (e) humic acid, 1000 mg/L; (f) fulvic acid-I, 3 mg/L; (g) fulvic acid-I, 30 mg/L; (h) fulvic acid-I, 300 mg/L; and (i) fulvic acid-I, 1000 mg/L (Inoue and Huang, 1987; reprinted with permission of Clay Minerals Society).

the presence of humic substances (Figure 19) were similar to those of the "allophanelike constituents" of Wada and Harward (1974) and Wada (1977, 1980) and to the "protoimogolite" allophane of Parfitt and Henmi (1980) and Farmer *et al.* (1980). HA at a concentration of 1000 mg/L almost completely inhibited the formation of allophanes or even of "allophanelike" materials or "protoimogolite" allophane and resulted in the formation of hydroxy-Al-humate complexes (Figure 19e). FA-I at a concentration of 1000 mg/L did not completely perturb the interaction of hydroxy Al ions with orthosilicic acid, but led to the formation of "protoimogolite" allophane (Figure 19i) complexed with FA. FA-II at the same concentration inhibited the formation of even "protoimogolite" allophane (not shown).

The accumulation of humus in the Dystrandept A_1 horizon may favor the formation of opaline silica, if the supply of silica is plentiful, and may retard the formation of "allophanelike" constituents, allophane, and imogolite, particularly the last two (Shoji and Masui, 1972; Tokashiki and Wada, 1975; Wada and Higashi, 1976; Higashi and Wada, 1977). The formation of imogolite from the solutions containing hydroxy Al ions and orthosilicic acid was strongly perturbed by humic substances (Inoue and Huang, 1984b). The data obtained by Inoue and Huang (1987) show that humic substances inhibit the formation of allophane, leading to the formation of hydroxy-Al-humic substances complexes and/or "protoimogolite" allophane. Al released from parent volcanic ash by weathering is strongly bound to humic substances, which limits the coprecipitation of Si and Al (Wada and Higiashi, 1976; Higashi and Wada, 1977) The findings of Inoue and Huang (1984a,b, 1985, 1987) substantiate the hypothesis that the accumulation of humus in the Dystrandepts A_1 and Spodosol A_2 and B_h horizons favors the formation of hydroxy-Al-humus complexes, "allophanelike" constituents, and/or opaline silica and may retard the formation of allophane and imogolite, which are formed in the Dystrandept B and (B) and Spodosol B_s horizons (Wada and Higashi, 1976; Higashi and Wada, 1977; Farmer *et al.*, 1980, 1983; Farmer, 1981, 1984; Anderson *et al.*, 1982).

VII. Impact of Aluminum Transformations on Soil and Environmental Sciences

Aluminum transformations have an enormous impact on soil and environmental sciences and human health, as summarized in Figure 20 and discussed below.

A. Weathering Reactions of Soil Minerals

Aluminum released to soil solution through chemical weathering of minerals can be adsorbed by interlayers of 2:1 expansible clay minerals (Jackson,

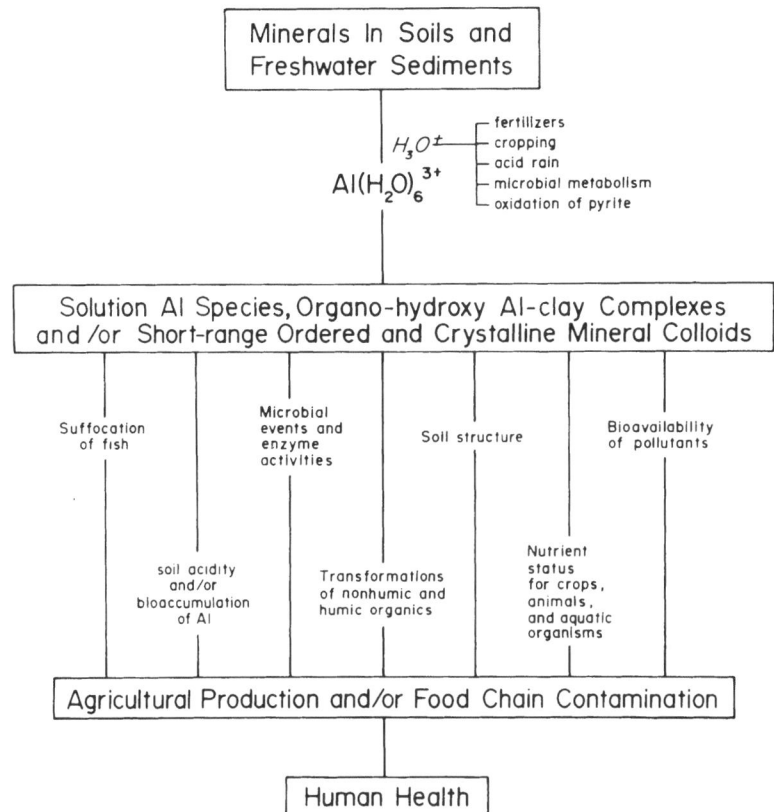

Figure 20. Impact of environmental Al chemistry on human health (Huang, 1987b; reprinted with permission of Soil Science Society of America).

1963b; Rich, 1968) or external surfaces of phyllosilicates (Huang and Kozak, 1970) or transformed to aluminum hydroxide polymorphs (Hsu, 1977) and short-range ordered precipitation products of aluminum (Huang and Violante, 1986). Among the three polymorphs of aluminum hydroxides, only gibbsite is common in soils (Hsu, 1977), and nordstrandite was found in some soils (Wall *et al.*, 1962).

Hydroxy-interlayered smectite and vermiculite occur in soils as a product of weathering reactions. These minerals may result from degradation of chlorite or deposition of hydroxy materials in the interlayers of expansible phyllosilicates. Pearson and Ensminger (1949), MacEwan (1950), Brown (1953), and Rich and Obenshain (1955) reported the presence of a 1.4-nm soil mineral that has properties similar to chlorite at room temperature. However, the 1.4-nm peak shifted toward a diffuse 1.0-nm d-spacing when the mineral was heated. Therefore, hydroxy-interlayered minerals do not fit the properties of either chlorite or vermiculite and smectite. Geographical distribution of this type of minerals is wide. They may occur in the soils of several soil taxonomic orders (Soil Survey Staff, 1975). The frequency of their occurrence is greatest in the Ultisols and the Alfisols. Hydroxy Al interlayering is frequently greatest in the surface horizon and decreases with depth. Frequency distribution and occurrence of Al-interlayered minerals in soils and sediments throughout the world have been reviewed by Rich (1968), Barnhisel (1977), and Bertsch and Barnhisel (1987).

B. Soil Physical Properties

Polynuclear Al species are known to exert important influences on soil physical properties. The swelling of Na montmorillonite is reduced by the formation of hydroxy Al and Fe interlayers (El Rayah and Rowell, 1973). Swelling is reduced to a greater extent with polynuclear Al species than with polynuclear Fe species. The removal of hydroxy interlayer materials has resulted in increased swelling of clays (Tamura, 1957). The tensile strength, liquid limit, shear-stress, and shrinking-swelling properties of montmorillonite clay systems may be significantly altered by hydroxy inter-layering and the accompanying changes in cation saturation (Dowdy and Larson, 1971; Gray and Schlocker, 1969; Davey and Low, 1971; Kidder and Reed, 1972).

The amount of sesquioxides extracted from a well-aggregated fraction of several prairie soils was considerably greater than that extracted from the poorly aggregated fraction (Weldon and Hide, 1942). Stable aggregates are heavily coated with aluminum and iron oxides, and the removal of these oxides destroys the aggregates (Kroth and Page, 1947). Saini *et al.* (1966) reported that the partial regression coefficient relating aluminum "oxides" with aggregation is 1.84 times as large as that for iron oxides. Jones and Uehara (1973) showed the presence of amorphous gel hull linkage between clay particles. Such amorphous gel hull is bound to include short-range ordered precipitation products of aluminum. El Swaify and Emerson (1975) reported the presence of aluminum hydroxide as a thin layer around illite particles. It is evident that sesquioxides act as cementing agents in the formation of aggregates. Aluminum hydroxides are more effective than iron oxides in maintaining the stability of soil aggregates except in fre-quently alternating oxidized and reduced soils. Well-crystallized alumi-num hydroxide may also act as a cementing agent in acidic conditions, but it may be negligible in its magnitude compared with short-range ordered precipitation products of aluminum. It has been suggested that organic matter may promote soil aggregation through the following linkage: clay–(Al, Fe)–organic matter–(Al, Fe)–clay (Edwards and Bremmer, 1967). Figure 21 shows an ultrathin section of a surface aggregate from a clay soil, cut next to a grass root (Emerson *et al.*, 1986). The pores between the

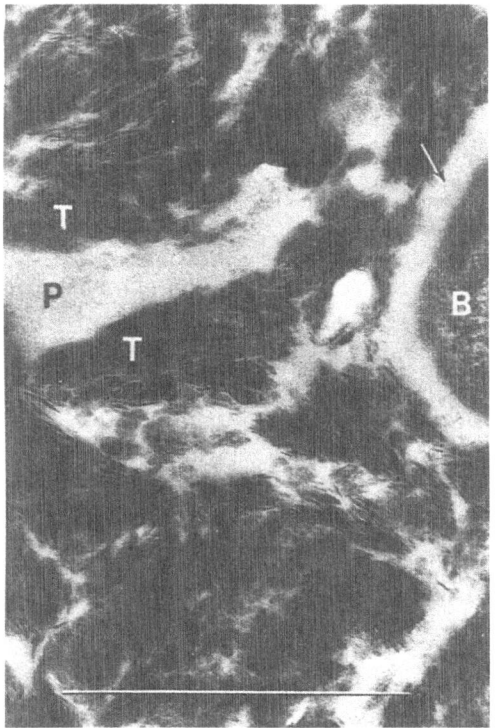

Figure 21. Ultrathin section of a surface soil treated with gluteraldehyde-La(OH)$_3$/O$_s$O$_4$ from a Wiesenboden (pellustert), a clay soil. Tactoids (T) are clearly visible with the intervening pores filled with acidic polysaccharides (P). A bacterium (B) is surrounded by extracellular polysaccharides (ECP) (arrow). Bar = 1μm (Emerson *et al.*, 1986; reprinted with permission of Soil Science Society of America).

tactoids are completely filled with a gel that consists of carbohydrate containing uronic acid groups. Aluminum ions may bridge uronic acid groups in the bonding polysaccharides.

C. Soil Acidity and Liming

Aluminum is released from soil minerals by chemical and biochemical weathering reactions. The hydrolysis and polymerization of Al release protons, leading to acidification of the environment. Released protons attack Al-bearing minerals, resulting in the dissolution of Al. The dissolved Al undergoes hydrolytic reactions and releases protons to soil solutions and natural waters. Therefore, generation of protons is perpetuated. Furthermore, this proton generation process can be accelerated by the presence of silica relics (pK$_1$ of Si(OH)$_4$ is 9.8) of silicates as a "proton sink." Consequently, the weathering of Al-bearing minerals, which can be enhanced by leaching, natural vegetation, farming, and other anthropogenic activities, would lead to acidification of soils and the associated environments. The hydroxy Al polymers adsorbed on the surfaces of soil minerals are important sources of soil acidity and are termed "third buffer range" acidity (Schwertmann and Jackson, 1963), pH-dependent, or titratable acidity (Clark, 1964; Colemen and Thomas, 1964; De Villiers and Jackson, 1967b). More recent data show that besides hydroxy Al polymers, organic acids that are chelated with the adsorbed Al in interlayers of clays can very substantially contribute to the total and titratable acidity (Goh and Huang, 1985). Furthermore, Al released from soil minerals may complex with a series of inorganic ligands, low-molecular-weight organic acids, and humic substances and form discrete colloidal precipitates in a wide range of particle size fractions of soils (Huang, 1980).

The interaction of the Al released into the soil solution with a series of organic components greatly modifies the nature of soil acidity. An increase in the contents of organic matter of soils results in a decrease in the levels of exchangeable Al and exchangeable acidity (McLean et al., 1965; Poinke and Corey, 1967; Thomas, 1975; Hoyt and Turner, 1975; Hoyt, 1977; Hargrove and Thomas, 1981). On the other hand, the contents of soil organic matter are positively correlated with the levels of titratable Al and acidity (Hoyt, 1977; Curtin et al., 1984). Aluminum is hydrolyzed to a great extent at low pH in the presence of organic matter, because the hydrolytic products react with carboxyl groups more intensively than with clays, and because the organic matter serves as a sink for H^+ ions produced by hydrolysis (Thomas and Hargrove, 1984).

Salt addition to sesquioxide-coated 2:1 expansible minerals appears to increase the hydrolysis of nonexchangeable Al (or Fe), resulting in an increase in H^+ ion concentration of the solution and to lower the pH (Coleman et al., 1964). Extensive hydrolysis of nonexchangeable Al occurred during the extraction of soils with neutral unbuffered salts (Kissel et al., 1971). Furthermore, the proportion of hydroxy Al that hydrolyzed was related to the stability of the hydroxy Al interlayers.

Titratable acidity arises from dissociation of weak-acid functional groups of soil organic matter (Hoyt, 1977) and deprotonation from hydroxy Al polymers associated with the internal (De Villiers and Jackson, 1967b) and external (Huang and Kozak, 1970) surfaces of phyllosilicate minerals, from short-range ordered aluminosilicate (allophanic) materials (Wada, 1977), and from ruptured surfaces of oxides and silicates (Huang, 1980). As liming programs are commonly designed to raise pH to 6.5–6.8 (Shoemaker et al., 1961), both the exchangeable and titratable components of acidity must be taken into account in assessing the lime needs of soils. Curtin et al. (1984) reported that KCl-exchangeable acidity was very low (0.3–2.2 mmol($+$) $\cdot$ kg^{-1}) in a group of 20 Saskatchewan soils (pH 5.0–5.8). Titratable acidity ranged from 9 to 191 mmol($+$) $\cdot$ kg^{-1} and constituted about

99% of the total acidity in the soils examined. Titratable acidity was highly correlated with organic carbon (r = 0.83) and with Al extracted using citrate-dithionite-bicarbonate (r = 0.95), potassium pyrophosphate (r = 0.92), and pH 4.8 ammonium acetate (r = 0.79). The combination of organic carbon and citrate-dithionite-bicarbonate extractable Al accounted for 96% of the variation in titratable acidity.

D. Formation of Humic Substances, Dynamics of Natural Organics and Enzymes, and Microbial Events

Alumina has catalytic effects on the formation of humic substances. In a respirometer experiment with clay minerals, the amounts of Al in the citrate-bicarbonate-dithionite extracts are proportional to the reduction of O_2 uptake after the treatment (Wang et al., 1978). The yields of humic substances formed from the catechol are significantly greater in the presence of aluminum oxides than in the blank (Wang et al., 1986). The Al on edges of kaolinite and short-range ordered aluminum oxides has the ability to catalyze the formation of humic substances from hydroquinone or pyrogallol (Wang and Huang, 1986, 1987).

The humic colloids in allophanic soils are believed to be stabilized by complexing with active Al groups or aluminum and iron oxides associated with allophane surfaces (Wada and Higashi, 1976). The greater stabilization of organic matter is related to a stronger interaction of active functional groups of organic colloids with allophane surfaces. The interaction may be direct adsorption to positive sites of allophanes or, more likely, to the formation of humus-Al, Fe, or Al, and Fe-allophane complexes.

The rate of the loss of C as CO_2 is slower from allophanic soils than from nonallophanic soils (Broadbent et al., 1964; Ino and Monsi, 1964; Jackman, 1964; Martin and Haider, 1986). Martin et al. (1982) followed the release of C as CO_2 over a 1-year period from a number of allophanic soils of southern Chile and the release from normal soils of California and Chile. Losses of C range from 143 to 243 mg/100 g^{-1} soil from nonallophanic soils containing about 16–26 g/kg^{-1} organic matter. Comparable losses from allophanic soils containing about 88–161 g/kg^{-1} organic matter range from 92 to 191 mg.

The C loss from a wide variety of ^{14}C-labeled readily degradable organic substrates in normal agricultural soils of California and Chile and allophanic soils of southern Chile was studied by Zunino et al. (1982). Over a 16-week incubation period, the presence of allophane in the soil or the addition of 50–160 g/kg^{-1} allophane to a very sandy soil reduced C losses by 16–73%. The least reduction occurred with glucose and the greatest with a bacterial polysaccharide in allophanic subsoil (Table 13). The losses of C from additions of more resistant polymer substrates such as plant lignins and fungal melanins are reduced to a much greater extent in the allophanic soils than losses in the normal agricultural soils. Influence of

Table 13. Decomposition of some ^{14}C-labeled organic substrates in the presence and absence of allophane clay

Soil	Percentage of added ^{14}C evolved as $^{14}CO_2$ in 16 weeks						
	Glucose	Polysaccharides	Wheat straw	*Chloella* protein	*Mucor* cells	Corn stalk lignin	*Aspergillus glaucus* melanin
Nonallophanic soils	77	75	60	67	55	17	5
Sandy loam + 16% allophane	58	56	37	41	36	—	—
Allophanic topsoils	56	44	34	38	39	4	2
Allophanic subsoil	56	21	25	26	34	—	—

Source: Compiled by Martin and Haider (1986) from the data of Martin *et al.* (1982) and Zunino *et al.* (1982).

aluminum-bearing minerals on the transformations of soil organic components through abiotic and biotic processes merits close attention.

The study of the interaction of enzymes with soil inorganic constituents has been mainly concerned with crystalline clay minerals (Burns, 1986). Ross (1983) reported that the original activities of α-amylase, β-amylase, and invertase were decreased by the presence of allophane to 57%, 12%, and 20%, respectively. The role of short-range ordered Al-bearing mineral colloids in influencing enzyme activities should receive more attention.

More recent evidence indicates that the utilization of organics by microbes seems to be modified when bound on metal oxide–coated clays, compared with clays homoionic to nonpolymeric cations (Stotzky, 1986). As the surface properties of these metal oxides vary with pH and ionic factors, the release of such bound organics may occur as the pH and ionic environments of natural microhabitats fluctuate. Once the organics have been released from the hydrous metal oxide–mineral complexes, they could be either degraded by soil microbes or complexed by mineral colloids. The role of short-range ordered Al hydrous oxides in the dynamics of transformations of organic compounds and in the ecology of microbes thus deserves increasing attention.

E. Transformations of Nutrients and Toxic Pollutants

Aluminum-bearing components in terrestrial and aquatic environments play a vital role in the transformations of nutrients and toxic substances, as stated below.

1. Nitrogen

a. Nitrate

Nitrate is weakly adsorbed on aluminum hydroxides by electrostatic attraction. Therefore nitrate can only be adsorbed by positively charged surfaces and is loosely held in the diffuse layer except in completely dehydrated systems. In most cases, nitrate is leached quite readily from soils and is held more weakly than chloride (Parfitt, 1978). Leached nitrate can be transported to aquatic environment. Soils such as volcanic ash soils that have positive sites on allophane retain nitrate more strongly than other soils and are thus able to prevent rapid leaching of nitrate.

b. Ammonium

Ammonium ions can be adsorbed on the negatively charged surfaces of Al components of soils and sediments. Furthermore, the adsorption of NH_4^+ ions by expansible phyllosilicates could also be affected by the deposition of hydroxy Al polymers on the interlamellar surfaces of the minerals (Huang, 1980).

c. Organic Nitrogen Compounds

Nitrogenous organic compounds (e.g., aspartic acid) can be adsorbed or coprecipitated with hydroxy Al compounds (Kwong and Huang, 1981). Such reactions could affect the transformations of nitrogen in the environment.

2. Phosphorus

Most research data show that phosphate adsorption is better associated with extractable Al than with iron (Parfitt, 1978). The affinity of phosphate for Al ions is strong enough to remove OH from an edge Al·OH. This type of reaction is referred to as a ligand exchange reaction. Phosphate can be adsorbed at the point of zero change (PZC) or on its alkaline side (Hsu, 1977). Ligand exchange reactions occur rapidly between exposed Al·OH groups and phosphate in solution. During the processes of weathering reactions, fresh aluminum hydroxide is continuously added to the system. The adsorbed phosphate may be occluded and become unavailable to plants if it is coated with another layer of aluminum hydroxides.

In the immediate vicinity of phosphate fertilizer particles, there are local conditions of low pH and high phosphate concentration, which may cause dissolution of clays and precipitation of aluminum phosphate. Basic aluminum phosphates may form in acid soils even at low phosphate concentrations. Formation of taranakite was observed by reaction of Al hydroxide with concentrated K^+ or NH_4^+ phosphate solution, but these compounds hydrolyzed to noncrystalline aluminum phosphate upon dilution and are slow-releasing phosphate sources rather than products of fixation.

Organic acids promote the formation and stability of short-range ordered precipitation products of Al with an accompanying high specific surface and thus help to maintain their high phosphate retention capacities (Kwong and Huang, 1979c). Nevertheless, organic acids above certain concentrations also promote aggregation of the precipitates (Kwong and Huang, 1979c; Violante and Huang, 1984). The relative influence of the exposure of surfaces as a result of structural distortion of precipitation products of Al and their aggregation caused by organic acids on the adsorption of anions merits further attention.

3. Potassium

Potassium ions can be specifically adsorbed by vermiculitic minerals, because the K^+ ion has a crystalline radius compatible to the size of adsorption sites. The selectivity of K^+ ions by vermiculite is increased by the deposition of hydroxy Al polymers on the interlamellar surfaces of the minerals. The mechanisms of the effect of hydroxy Al interlayer on the K/Ca exchange selectivity may proceed through (1) propping effect, (2) preferential occupation of Ca adsorbing sites, and (3) retarding effect on the entry of the more hydrated Ca ions (Figure 22). Propping effect would

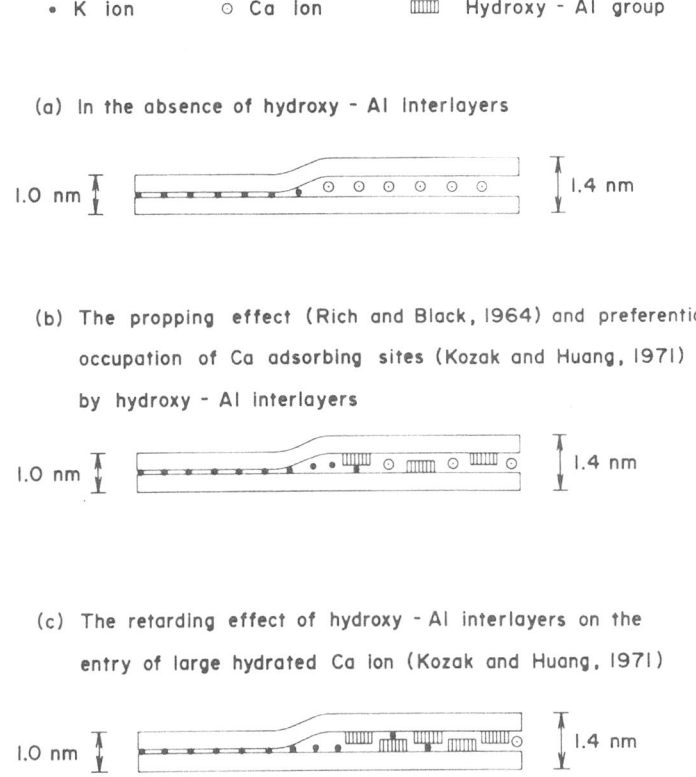

Figure 22. Structural models of the mechanisms of the effects of hydroxy Al interlayers on the K/Ca exchange selectivity of 2:1 phyllosilicates which have wedge zones and rather limited interlayer expansibility (Huang, 1987a; reprinted with permission of Pergamon Books Ltd.).

prevent K fixation and increase K selectivity by permitting its entry to vermiculite and weathered mica. Potassium ions are thus selected at wedge sites, whereas large hydrated ions such as Ca and Mg are excluded. The preferential occupation by hydroxy Al on exchange sites that normally would adsorb the more hydrated Ca ion would result in the concentration of K-retaining sites in interlayers. The hydroxy interlayers also create a retarding effect on the entry of the more hydrated Ca ions, thus effectively increasing the selectivity of the minerals toward the less hydrated K ions. The formation of hydroxy Al–clay complexes would thus affect the transformations of K^+ ions in the environment.

4. Sulfur

Soils with large amounts of hydrous oxides of Al and Fe retain substantial amounts of sulfate (Parfitt, 1978). Therefore, tropical soils and volcanic ash soils strongly adsorbed sulfate. Sulfate adsorption has been shown to be correlated with extractable Al rather than Fe, although both hydrous Al and Fe oxides strongly adsorb sulfate.

Aluminum is more important than Fe in sulfate adsorption near to the saturation level. Loesses of sulfate by leaching are more significant in soils that are low in hydrous oxides of Al and Fe.

The mechanism of ligand exchange of $Al \cdot OH_2$ or $Al \cdot OH$ groups by sulfate ions can take place only on the acidic side of PZC (Hsu, 1977). The adsorption of sulfate by Al hydroxides greatly increases with decreasing pH.

5. Alkaline Earth and Heavy Metals

Alkaline earth (Ca, Mg, Ba) and heavy metals (Cu, Zn, Mn, Fe, Ni, Cr, Hg, Cd, Pb, Sn, U, V, Co, Ag) either are essential elements to plant growth and animal and human nutrition or are toxic environmental pollutants (Underwood, 1977).

Hydrous oxides of Al specifically adsorb alkaline earth metals such as Ba, Ca, Sr, and Mg and heavy metals such as Cu, Zn, Mn, and Hg (Huang, 1980; Kinniburgh and Jackson, 1981). Compared with aged oxides, fresh oxides have a much higher adsorptive capacity for Zn. Generally, Zn adsorption capacity is directly related to cation exchange capacity of clays and hydrous oxides. The surface reactivity of hydrous oxides of Al toward Ca is greatly enhanced by the presence of low-molecular-weight organic acids commonly present in soils during the formation of these oxides (Kwong and Huang, 1979c). Many organic acids prevalent in terrestrial and aquatic environments are very effective in perturbing the hydrolytic reaction of Al and promoting the formation of short-range ordered Al oxides. The influence of these organic acids on the surface reactivity of Al oxides toward alkaline earth and heavy metals warrants further studies.

6. Arsenic

Arsenic (V) is probably the dominant oxidation state of As in soils (Davies, 1980) and sediments. The adsorption of arsenate by soils and sediments is controlled largely by the hydrous oxides of Al and Fe. Arsenate could be desorbed from soils by leaching with phosphate, indicating that arsenate, like phosphate, is adsorbed on hydrous oxides by a ligand exchange mechanism.

7. Selenium

Selenites represent the most important form of Se in many soils (Davies, 1980). Selenite is immobilized by sesquioxides in acid soils, and its leaching is increased by additions of sulfate. This indicates that both selenite and sulfate are adsorbed on the same sites.

8. Boron

Much adsorption of B is attributed to hydroxy Al and Fe compounds present as coatings on surfaces of clays (Barber, 1984). Boron retention by Al compounds is far greater than that effected by analogous iron structures. Maximum adsorption of B by soils is observed at pH 9, which is close to the pK value of the formation of $B(OH)_4^-$ from $B(OH)_3$.

9. Molybdenum

Molybdate is adsorbed on surfaces of Al and Fe oxides by a ligand exchange reaction with exposed Al $\cdot$ OH and Fe $\cdot$ OH groups (Parfitt, 1978). The adsorption of molybdate increases from low pH values to a maximum at pH 4. Molybdate adsorption on different soils paralleled phosphate and sulfate adsorption.

10. Fluorine, Chlorine, Iodine

The reaction of fluoride with Al oxides and Al-bearing clay minerals involves the release of OH to solution. At low concentrations, the mechanism appears to involve exchange with the OH of Al hydroxides. At high concentrations, the complexation and release of Al is the mechanism of the reaction (Huang and Jackson, 1965).

Iodide is adsorbed by Al and Fe hydroxides. Sorption of I^- on sesquioxides is important only at pH < 6.

Chloride is adsorbed by electrostatic attraction to Al $\cdot$ $OH_2^{0.5+}$ sites that occur at low pH (Parfitt, 1978). Adsorbed Cl^- is exchangeable with NO_3^- and could be desorbed by leaching with water.

11. Pesticides

Hydrous oxides of Al and noncrystalline aluminosilicates, allophane, have large surface areas. These materials carry both positive and negative charges on their surfaces. The interaction of cationic and anionic pesticides with such materials (Green, 1974) merits further study. Besides ionic pesticides, nonionic compounds that are organic bases such as s-triazine, when protonated, can be adsorbed by soil colloids. The proton-donating capability of clay surfaces is significantly enhanced when Al^{3+} is the saturating cation. The role of Al and its oxides in the adsorption and persistence of s-triazine herbicides such as atrazine (Huang et al., 1984) has been demonstrated. Short-range ordered oxides of Al and Fe present in a series of particle size fractions of soils deserve close attention in defining soil components governing the degree of adsorptivity of pesticides by soils.

F. Ecological and Health Effects

The extent of the release of Al from minerals through acidification of the environment has increased with time, population growth, intensification of agriculture, and industrialization. The Al^{3+} ion of soils damages the roots of some common crops such as barley and corn (*Zea mays*). The sensitivity of wheat to Al is conditioned by a single recessive gene. The effects of Al are a result of an interference in DNA replication in cell division, and the uptake by plants is usually excluded metabolically (Clarkson, 1969). With some crop plants, the accumulation of Al is mainly kept in the roots, with relatively little transported to the above-ground portions. Some plants develop resistance to Al intake and toxicity through Al chelation, precipitation, and other adaptations, and thus Al is forwarded upward into the food chain. Aluminum accumulator food plants include tea (*Thea sinensis* L.), spices such as oregano, and asparagus (*Asparagus officinalis*). It appears important to lower plant-available Al^{3+} in agricultural soils by the addition of ground limestone, with the provision of adequate supplies of Mg. Aluminum availability to some plants is repressed as the soil pH is raised up to pH 6.5 in some regions. This practice affords some protection of plants and animals from the toxicity of excess Al. Adequately liming acid soils thus provides an ecology of health on land.

Acid rains increase Al^{3+} levels in lake waters through the interaction of protons with Al-bearing minerals in suspended and bottom sediments. The enhanced Al^{3+} levels in lake waters cause Al gel accumulation in the neutral gills of fish, and suffocation results (Cronan and Schofield, 1979). Information concerning the toxicity of Al compounds of freshwater fish can be summarized in three categories: (1) lethal concentrations and causes of mortality, (2) subchronic concentrations and their effects, and (3) genotoxic or embryotoxic concentrations and their effects. The ecological effects of Al compounds on freshwater fish were thoroughly reviewed by Burrows (1977).

The human body burden is normally only 30 mg of Al (Alfrey, 1980). Yet the daily intake and excretion varies from 20 to 40 mg of Al in different countries (Underwood, 1977). Effective mechanisms for exclusion of Al and its potential toxicity from the normal human body evidently exist.

Acid soil–related diseases of the central nervous systems, however, have been identified in Guam and Japan (Yase, 1980). High levels of acid soil Al^{3+} and food Al^{3+} together with low Mg/Ca ratios are associated with amyotrophic lateral sclerosis (ALS) and Parkinsonism dementia (PD), with Al-enriched hydroxyapatite deposits in the spinal nerve region and brain of ALS/PD cases. The Al^{3+} of acid soils is particularly damaging in the food chain if Mg is not adequate in supply to maintain its function in animal hexokinase or "ATPase" necessary for glucose energy release. The Al-induced metabolic aberrations interfere with phosphorylation of sugars and impair P translocation (Underwood, 1977). Replacement of Mg^{2+} from hexokinase by Al^{3+} may cause potentially lethal "grass tetany" in cattle.

Water that receives Al treatment to lower its F content, when used in kidney dialysis, induced encephalopathology. Nondispersive X-ray spectroscopic analysis of the amyloid plaques of the hippocampus portion of brain associated with human senile dementia of the Alzheimer type reveals a high Al concentration with some Si and Mg.

Aluminum ions react with various inorganic and organic constituents common in soils and the associated environments (Huang and Violante, 1986). However, insufficient information is available with respect to Al speciation relative to absorption, transport, distribution, storage, excretion, and metabolic interaction in any living system. Problems involving identification of a variety of aluminum species and characterization of their chemical and biological effects merit further research.

VIII. Summary and Conclusions

Aluminum is released from Al-bearing minerals to soil solutions and natural waters through the action of protons and complexing ions during chemical and biochemical weathering processes. The release of Al from mineral structure to soil solutions and natural waters leads to the hydrolysis and/or polymerization of Al and the subsequent formation of a series of solution Al species, organohydroxy Al–clay complexes, Al hydroxides and oxyhydroxides, and ill-defined short-range ordered mineral colloids.

Clay minerals can be considered as solid-state ions, and their particle size and structural and surface properties have an enormous influence on Al transformations. The ability of 2:1 expansible clays to fix polynuclear Al has been proposed as a reason for the absence of gibbsite in temperate soils. In addition to the formation of Al interlayered clays displaying a stable 1.4-nm d-spacing, direct magic-angle spinning NMR evidence indicates that Al_{13} polynuclear can be deposited in the interlayer space of the 2:1 clay minerals with a stable 1.8-nm d-spacing. However, the clay properties that affect the formation of Al_{13} interlayers remain obscure. Little is known about the mechanisms of the polymerization reactions of Al at the mineral water interface and their importance to the formation of polynuclear species in soils.

The nature and concentrations of inorganic ligands play an important role in influencing the hydrolysis and polymerization of Al. With anions that do not have strong affinity for Al, polynuclear Al ions continue to hydrolyze and polymerize unless a high concentration of the counteranion is present. In the presence of inorganic ligands that have strong affinity for Al, the further hydrolysis of polynuclear Al species is prevented or at least retarded. These counterpolyvalent anions tend to link polynuclear Al species together but in distorted arrangements for steric reasons. Most of such precipitation products are thus short-range ordered mineral colloids and different from Al hydroxide polymorphs. These ligands, with strong affinity for Al, can also promote physical flocculation, resulting from reduction in electrostatic repulsion between polynuclear Al ions.

Organic components are an integral part of soils and the associated environments. Low-molecular-weight biochemicals such as organic acids are being constantly introduced to the environment through natural vegetation and through farming and other anthropogenic activities. FAs and HAs account for the bulk of organic components of soils and sediments. Therefore, transformations of Al in soils and aquatic sediments take place in the presence of organics. These organic ligands disrupt the hydroxyl bridging mechanism which is indispensable for the formation of crystalline Al hydroxides. Because of steric factors, these perturbing organic ligands that occupy the coordination sites of Al also distort the arrangement of the platelets normally found in crystalline Al hydroxides, leading to the formation of short-range ordered precipitation products of Al. Consequently, it is logical that substantial concentrations of crystalline Al hydroxides are absent in temperate soils rich in organic matter. The research data also indicate that organic acids can perturb the formation of hydroxy-Al-montmorillonite complexes even in the acidic pH condition, which is most conducive to Al interlayering.

Organic acids can also influence the formation of Al hydroxide polymorphs through modification of the kinetics of the crystallization of precipitation products of Al. The organo-aluminopolymer associations also perturb the interaction of hydroxy Al ions with silicic acid and inhibit the formation of imogolite and allophane, leading to the accumulation of ill-defined aluminosilicates that have low SiO_2/Al_2O_3 ratios and contain considerable amounts of organic ligands.

Certain organic and inorganic ligands not only promote the formation of pseudoboehmites over Al hydroxides but also stabilize them. These results provide the interpretations for the genesis of noncrystalline alumina of boehmite characteristics in certain tropical soils and of some bauxites that consist of submicroscopic boehmitelike particles very similar to pseudoboehmites.

Complexation of Al by a series of organic and inorganic ligands may have a significant influence on the chemistry of Al in soil solutions and natural waters. The ecological significance of these Al species and the impact on human health merit close attention.

Besides affecting soil acidity, microbial events, enzyme activities, transformations of organic matter, soil structure, and the status of nutrients for plant and animal production, the transformation of Al can also affect the bioavailability of environmental pollutants such as heavy metals and biocides, and the subsequent food chain contamination in terrestrial systems. The hydrolytic products of Al can also influence the transformations, pathways, accumulation, and ecological effects of pollutants in freshwater environments. Therefore, the research on the mineralogy and environmental chemistry of Al is essential in the pursuits of environmental toxicology and in the studies pertaining to human health.

Acknowledgments

Financial support (Grant No. A2383—Huang) from the Natural Sciences and Engineering Research Council of Canada is appreciated.

References

Adams, F., and Z. Rawajfih. 1977. Basaluminite and alunite: A possible cause of sulfate retention by acid soils. *Soil Sci. Soc. Am. J.* 41:686–692.

Akitt, J.W., and A. Farthing. 1978. New ^{27}Al NMR studies of the hydrolysis of the aluminum (III) cation. *J. Maq. Res.* 32:345–352.

Akitt, J.W., and A. Farthing. 1981a. Aluminum-27 nuclear magnetic resonance studies of the hydrolysis of aluminum (III). 2. Gel-permeation chromatography. *J. Chem. Soc. Dalton* 1606–1608.

Akitt, J.W., and A. Farthing. 1981b. Aluminum-27 nuclear magnetic resonance studies of the hydrolysis of aluminum (III). 3. Stopped-flow kinetics studies. *J. Chem. Soc. Dalton* 1609–1614.

Akitt, J.W., and A. Farthing. 1981c. Aluminum-27 nuclear magnetic resonance studies of the hydrolysis of aluminum (III). 4. Hydrolysis using sodium carbonate. *J. Chem. Soc. Dalton* 1617–1623.

Akitt, J.W., and A. Farthing. 1981d. Aluminum-27 nuclear magnetic resonance studies of the hydrolysis of aluminum (III). 5. Slow hydrolysis using aluminum metal. *J. Chem. Soc. Dalton* 1626–1628.

Akitt, J.W., N.N. Greenwood, B.L. Khandelwal, and G.D. Lester. 1972. ^{27}Al nuclear magnetic resonance studies of the hydrolysis and polymerization of the hexa-aquaaluminum (III) cation. *J. Chem. Soc. Dalton* 604–610.

Aldcroft, D., G.C. Bye, and C.A. Hughes. 1969. Crystallization process in aluminum hydroxide gels. IV. Factors influencing the formation of the crystalline trihydroxides. *J. Appl. Chem.* 19:167–172.

Alfrey, A.C. 1980. Aluminum metabolism in uremia. In: L. Liss (ed.) *Aluminum Neurotoxicity*. Pathotox Publishers, Park Forest South, IL, pp. 43–53.

Anderson, H.A., M.L. Berrow, V.C. Farmer, A. Hepburn, J.D. Russell, and A.D. Walker. 1982. A reassessment of podzol formation processes. *J. Soil Sci.* 33:125–136.

Aveston, J. 1965. Hydrolysis of the aluminum ion: Ultracentrifugation and acidity measurements. *J. Chem. Soc.* 4438–4443.

Baes, C.F. Jr., and R.E. Mesmer. 1976. *The Hydrolysis of Cations*. John Wiley, New York, pp. 112–123.

Barber, S.A. 1984. *Soil Nutrient Bioavailability*. John Wiley, New York.

Bardossy, G., and J.L. White. 1979. Carbonate inhibits the crystallization of aluminum hydroxide in Bauxite. *Science* 203:355–356.

Barnhisel, R.I. 1965. *The Formation and Stability of Aluminum Interlayers in Clays*. Ph.D. Thesis, Virginia Polytechnic Institute, Blacksburg.

Barnhisel, R.I. 1969. Changes in specific surface areas of clays treated with hydroxy-aluminum. *Soil Sci.* 107:126–130.

Barnhisel, R.I. 1977. Chlorites and hydroxy interlayered vermiculite and smectite. In: J.B. Dixon and S.B. Weed (eds.) *Minerals in Soil Environments*. Soil Science Society of America, Madison, WI, pp. 331–356.

Barnhisel, R.I., and P.M. Bertsch. 1982. Aluminum. In: A.L. Page (ed.) *Methods of Soil Analysis*, Part 2, 2d ed. Agronomy No. 9. American Society of Agronomy, Madison, WI, pp. 275–300.

Barnhisel, R.I., and C.I. Rich. 1963. Gibbsite formation from aluminum-interlayers in montmorillonite. *Soil Sci. Soc. Am. Proc.* 27:632–635.

Barnhisel, R.I., and C.I. Rich. 1965. Gibbsite, bayerite and nordstrandite formation as affected by anions, pH, and mineral surfaces. *Soil. Sci. Soc. Am. Proc.* 29:531–534.

Bersillon, J.L., P.H. Hsu, and Fiessinger. 1980. Characterization of hydroxy-aluminum solutions. *Soil Sci. Soc. Am. J.* 44:630–634.

Berthelin, J., and G. Belgy. 1979. Microbial degradation of phyllosilicates during simulated podzolization. *Geoderma* 21:297–310.

Bertsch, P.M. 1988. Aqueous polynuclear species. In: G. Sposito (ed.) *The Environmental Chemistry of Aluminum*. CRC Press, Boca Raton, FL.

Bertsch, P.M., and R.I. Barnhisel. 1987. Chlorites and hydroxy-interlayered vermiculite and smectite. In: J.B. Dixon and S.B. Weed (eds.) *Minerals in Soil Environment*, 2d ed. Soil Science Society of America, Madison, WI.

Bertsch, P.M., G.W. Thomas, and R.I. Barnhisel. 1986. Characterization of hydroxy aluminum solutions by aluminum-27 nuclear magnetic resonance spectroscopy. *Soil Sci. Soc. Am. J.* 50:825–830.

Bloomfield, C. 1964. Organic matter and soil dynamics. In: E.G. Hallsworth and D.V. Crawford (eds.) *Experimental Pedology*. Butterworth, London, pp. 257–266.

Bokhari, U.G., D.C. Coleman, and A. Rubink. 1979. Chemistry of root exudates and rhizosphere soils of prairie plants. *Can. J. Bot.* 57:1473–1477.

Bolton, H.C. 1880. Action of organic acids on minerals. *Mineral. Mag.* 19:1–8.

Bolton, H.C. 1882. Application of organic acids to the examination of minerals. *Proc. Am. Assoc. Adv. Sci.* 31:3–7

Bottero, J.Y., J.M. Cases, F. Fiessinger, and J.E. Poirier. 1980. Studies of hydrolyzed aluminum chloride solutions. I. Nature of aluminum species and composition of aqueous solutions. *J. Phys. Chem.* 84:2933–2939.

Brady, N.C. 1974. *The Nature and Properties of Soils*, 8th Ed. Macmillan, New York, pp. 19–39.

Broadbent, F.E., R.N. Jackson, and J. McNicoll. 1964. Mineralization of C and N in some New Zealand allophanic soils. *Soil Sci.* 98:118–132.

Brosset, C. 1952. On the reactions of the aluminum ion with water. *Acta Chem. Scand.* 6:910–940.

Brosset, C., G. Biederman, and L.G. Sillen. 1954. Studies on the hydrolysis of metal ions. XI. The aluminium ion, Al^{3+}. *Acta Chem. Scand.* 8:1917–1926.

Brown, G. 1953. The dioctahedral analogue of vermiculite. *Clay Miner. Bull.* 2:64–70.

Bruckert, S. 1970a. Influence des composés organiques solubles sur la pédogènèse en milieu acide. I. Etude de terrain. *Ann. Agron.* 21:421–452.

Bruckert, S. 1970b. Influence des composés organiques solubles sur la pédogènèse en milieu acide. II. Experiences de laboratoire. *Ann. Agron.* 21:725–757.

Bruckert, S., F. Toutain, J. Tchicaya, and F. Facquin. 1971. Influence des pluviolessivats de hetre et de pin sylvestre sur les processus d'humification. *Oecol. Plant.* 6:329–339.

Brydon, J.E., and H. Kodama. 1966. The nature of aluminum hydroxide-montmorillonite complexes. *Am. Mineral.* 51:875–889.

Buffle, J., N. Parthasarathy, and W. Haerdi. 1985. Importance of speciation methods in analytical control of water treatment processes with application to fluoride removal from waste waters. *Water Res.* 19:7–23.

Bruns, R.G. 1986. Interaction of enzymes with soil minerals and organic colloids. In: P.M. Huang and M. Schnitzer (eds.). *Interactions of Soil Minerals with Natural Organics and Microbes.* SSSA Special Publication No. 17. Soil Science Society of America, Madison, WI, pp. 429–451.

Burrows, W.D. 1977. Aquatic aluminum chemistry, toxicology, and environmental prevalence. *CRC Crit. Rev. Environ. Control* 7:167–216.

Campbell, A.S., A.W. Young, L.G. Livingstone, M.A. Wilson, and T.W. Walker. 1977. Characterization of poorly-ordered alumino-silicates in a vitric Andosol from New Zealand. *Soil Sci.* 123:362–368.

Campbell, P.G.C., R. Bougie, A. Tessier, and J.P. Villeneuve. 1985. Aluminum speciation in surface waters on the Canadian Pre-Cambrian shield. *Verh. Int. Verein. Limnol.* 22:371–375.

Carreira, L.A., V.A. Maroni, J.W. Swaine, and R.C. Plumb. 1966. Raman and infrared spectra and structures of the aluminate ions. *J. Chem. Phys.* 45:2216–2220.

Carstea, D.D. 1968. Formation of hydroxy-Al and -Fe interlayers in montmorillonite and vermiculite: Influence of particle size and temperature. *Clays Clay Miner.* 16:231–238.

Casalicchio, G., and N. Rossi. 1970. Investigations on composition of the soil lipid fraction. I. Determination of some free organic acids. *Agrochimica* 14:505–515.

Chesworth, W. 1972. The stability of gibbsite and boehmite at the surface of the earth. *Clays Clay Miner.* 20:369–374.

Clark, J.S. 1964. Aluminum and iron fixation in relation to exchangeable hydrogen in soils. *Soil Sci.* 98:302–306.

Clarkson, D.T. 1969. Metabolic aspects of aluminum toxicity and some possible mechanisms of resistance. In: I.H. Rorison (ed.), *Ecological Aspects of Mineral Nutrition of Plants.* Blackwell Scientific, Oxford, U.K., pp. 381–397.

Clément, A. 1977. Comparaison de la nutrition minerale de Pinus nigra Nigricans et de Picea excelsa Link en sol très carbonaté, carbonaté et décarbonaté: Incidence sur le métabolisme des anions mineraux et organiques. *Ann. Sci. Forest.* 34:293–309.

Coleman, N.T., and G.W. Thomas. 1964. Buffer curves of acid clays as affected by the presence of ferric iron and aluminum. *Soil Sci. Soc. Am. Proc.* 28:187–190.

Coleman, N.T., G.W. Thomas, F.H. Le Roux, and G. Bredell. 1964. Salt-exchangeable and titratable acidity in bentonite-sesquioxide mixtures. *Soil Sci. Soc. Am. Proc.* 28:35–37.

Cotton, F.A., and G. Wilkinson. 1980. *Advanced Inorganic Chemistry*, 4th Ed. Interscience Publishers, John Wiley, New York.

Coulson, C.B., R.I. Davies, and D.A. Lewis. 1960. Polyphenols in plant, humus and soils. I. Polyphenols of leaves, litter and superficial humus from mull and mor sites. *J. Soil Sci.* 11:20–29.

Cronan, C.S., and C.L. Schofield. 1979. Aluminum leaching response to acid precipitating effects on high-elevation watersheds in the Northeast. *Science* 204:304–306.

Curtin, D., H.P.W. Rostad, and P.M. Huang. 1984. Soil acidity in relation to soil properties and lime requirement. *Can. J. Soil Sci.* 64:545–554.

Das Sarma, B. 1956. The structure of copper monoglutamate. *J. Am. Chem. Soc.* 78:892–894.

Davey, B.G., and P.F. Low. 1971. Physico-chemical properties of sols and gels of Na-montmorillonite with and without adsorbed hydrous aluminum oxide. *Soil Sci. Soc. Am. Proc.* 35:230–236.

Davies, B.E. 1980. *Applied Soil Trace Elements.* John Wiley, New York.

Davies, R.J. 1971. Relation of polyphenols to decomposition of organic matter and to pedogenetic processes. *Soil Sci.* 111:80–85.

Davis,C.E., and V.G. Hill. 1974. Occurrence of nordstrandite and its possible significance in Jamaica bauxites. *Travaux* 11:61–70.

De Hek, H., R.J. Stol, and P.L. DeBruyn. 1978. Hydrolysis-precipitation studies of aluminum (III) solutions. 3. The role of the sulfate ion. *J. Colloid Int. Sci.* 64:72–89.

Desai, A.D., and S.T. Rao. 1957. Injurious effect of green manure on rice under ill-drained conditions. *J. Indian Soc. Soil Sci.* 5:147–153.

De Villiers, J.M. 1969. Pedosesquioxides—composition and colloidal interactions in soil genesis during the Quaternary. *Soil Sci.* 107:454–461.

De Villiers, J.M., and M.L. Jackson. 1967a. Cation exchange capacity variations with pH in soil clays. *Soil Sci. Soc. Am. Proc.* 31:473–476.

De Villiers, J.M., and M.L. Jackson. 1967b. Aluminous chlorite origin of pH-dependent cation exchange capacity variation. *Soil Sci. Soc. Am. Proc.* 31:614–619.

Dowdy, R.H., and W.E. Larson. 1971. Tensile strength of montmorillonite as a function of saturating cation and water content. *Soil Sci. Soc. Am. Proc.* 35:1010–1014.

Duff, R.B., and D.M. Webley. 1959. 2-Ketogluconic acid as a natural chelator produced by soil bacteria. *Chem. Ind.* 1376–1377.

Duff, R.B., D.M. Webley, and R.O. Scott. 1963. Solubilization of minerals and related materials by 2-ketogluconic acid–producing bacteria. *Soil Sci.* 95:105–114.

Edwards, A.P., and J.M. Bremner. 1967. Microaggregates in soils. *J. Soil. Sci.* 18:64–73.

El Rayah, H.M.E., and D.L. Rowell. 1973. The influence of iron and aluminum hydroxides on the swelling of Na-montmorillonite and the permeability of a Na-soil. *J. Soil Sci.* 24:137–144.

El Swaify, S.A., and W.W. Emerson. 1975. Changes in the physical properties of soil clays due to precipitated aluminum and iron hydroxides. I. Swelling and aggregate stability after drying. *Soil Sci. Soc. Am. Proc.* 39:1056–1063.

Emerson, W.W., R.C. Foster, and J.M. Oades. 1986. Organo-mineral complexes in relation to soil aggregation and structure. In: P.M. Huang and M. Schnitzer (eds.) *Interactions of Soil Minerals with Natural Organics and Microbes.* SSSA Special Publication No. 17. Soil Science Society of America, Madison, WI, pp. 521–548.

Farmer, V.C. 1981. Possible roles of a mobile hydroxyaluminium orthosilicate complex (proto-imogolite) and other hydroxyaluminium and hydroxy-iron species in podzolization. In: *Migrations Organo-Minérales dans les Sols Tempérés.* Colloques Internationaux du CNRS No. 303, pp. 275–279.

Farmer, V.C. 1984. Distribution of allophane and organic matter in podzol B horizons: Reply to Buurman and Reeuwijk. *J. Soil Sci.* 3:453–458.

Farmer, V.C., and A.R. Fraser. 1979. Synthetic imogolite, a tubular hydroxyaluminium silicate. In: M.M. Mortland and V.C. Farmer (eds.) *Proc. 6th Int. Clay Conf. (Oxford)*, pp. 547–553.

Farmer, V.C., A.R. Fraser, J.D. Russel, and N. Yoshinaga. 1977. Recognition of imogolite structures in allophanic clays by infrared spectroscopy. *Clay Miner.* 12:55–57.

Farmer, V.C., A.R. Fraser, J.M. Tait, F. Palmieri, P. Violante, M. Nakai, and N. Yoshinaga. 1978. Imogolite and protoimogolite in an Italian soil developed on volcanic ash. *Clay Miner.* 13:271–274.

Farmer, V.C., A.R. Fraser, and J.M. Tait. 1979. Characterization of the chemical structures of natural and synthetic aluminosilicate gels and sols by infrared spectroscopy. *Geochim. Cosmochim. Acta* 43:1417–1420.

Farmer, V.C., J.D. Russell, and M.L. Berrow. 1980. Imogolite and protoimogolite allophane in spodic horizons: Evidence for a mobile aluminium silicate complex in podzol formation. *J. Soil Sci.* 31:673–684.

Farmer, V.C., J.D. Russell, and B.F.L. Smith. 1983. Extraction of inorganic forms of translocated Al, Fe and Si from a podzol Bs horizon. *J. Soil Sci.* 34:571–576.

Fitzpatrick, R.W., and U. Schwertmann. 1982. Al-substituted goethite—an indicator of pedogenic and other weathering environments in South Africa. *Geoderma* 27:335–347.

Flaig, W. 1982. Dynamics of organic matter decomposition in soils. In: *Nonsymbiotic Nitrogen Fixation and Organic Matter in the Tropics*, 12th Int. Cong. Soil Sci. New Delhi, 1:115–124.

Flaig, W., H. Beutelspacher, and E. Rietz. 1975. Chemical composition and physical properties of humic substances. In: J.E. Gieseking (ed.) *Soil Components*, Vol. 1: *Organic Components*. Springer-Verlag, New York, pp. 1–211.

Förstner, U. 1981. Metal transfer between solid and aqueous phases. In: U. Förstner and G.T.W. Wittmann (eds.) *Metal Pollution in the Aquatic Environment*. Springer-Verlag, Heidelberg, pp. 197–270.

Frink, C.R. 1965. Characteristics of aluminum interlayers in soil clays. *Soil Sci. Soc. Am. Proc.* 29:379–382.

Frink, C.R., and M. Peech. 1963. Hydrolysis of the aluminum ion in dilute aqueous solution. *Inorg. Chem.* 2:473–478.

Frink, C.R., and B.L. Sawhney. 1967. Neutralization of dilute aqueous aluminum salt solutions. *Soil Sci.* 103:144–148.

Fripiat, J.J., F. van Cauwelaert, and H. Bosmans. 1965. Structure of aluminum cations in aqueous solutions. *J. Phys. Chem.* 69:2458–2461.

Goh, T.B., and P.M. Huang. 1984. Formation of hydroxy-Al-montmorillonite complexes as influenced by citric acid. *Can. J. Soil Sci.* 64:411–421.

Goh, T.B., and P.M. Huang. 1985. Changes in the thermal stability and acidic characteristics of hydroxy-Al-montmorillonite complexes formed in the presence of citric acid. *Can. J. Soil Sci.* 65:519–522.

Goh, T.B., and P.M. Huang. 1986. Influence of citric and tannic acids on hydroxy-Al interlayering in montmorillonte. *Clays Clay Miner.* 34:37–44.

Gomah, A.M., and N.N.C. Sakhar. 1972. Heavy metal chelates in herbage and soil profiles from upland sites in Wales. *Rep. Welsh Soils Discussion Group* 13:17–39.

Gray, D.H., and J. Schlocker. 1969. Electrochemical alteration of clay soils. *Clays Clay Miner.* 17:309–322.

Green, R.E. 1974. Pesticide-clay-water interactions. In: W.D. Guenzi (ed.). *Pesticides in Soil and Water.* Soil Science Society of America, Madison, WI, pp. 3–37.

Greenland, D.J. 1965a. Interaction between clays and organic compounds in soils. I. Mechanisms of interaction between clays and defined organic compounds. *Soils Fertilizers* 28:415–425.

Greenland, D.J. 1965b. Interaction between clays and organic compounds in soils. II. Adsorption of soil organic compounds and its effect on soil properties. *Soils Fertilizers* 28:521–532.

Hargrove, W.L., and G.W. Thomas. 1981. Effect of organic matter on exchangeable aluminum and plant growth in acid soils. In: R.H. Dowdy, J.A. Ryan, V.V. Volk, and D.E. Baker (eds.), *Chemistry in the Soil Environment.* Soil Science Society America, Madison, WI, pp. 151–165.

Hayes, M.H.B., and F.L. Himes. 1986. Nature and properties of humus mineral complexes. In: P.M. Huang and M. Schnitzer (eds.) *Interactions of Soil Minerals with Natural Organics and Microbes.* SSSA Special Publication No. 17. Soil Science Society of America, Madison, WI, pp. 103–158.

Hem, J.D. 1968. Graphical methods for studies of aqueous aluminum hydroxide, fluoride, and sulfate complexes. U.S. Geol. Surv. Water Supply Paper 1827B, 33 pp.

Hem, J.D., and C.E. Roberson. 1967. Form and stability of aluminum hydroxide complexes in dilute solution. U.S. Geol. Surv. Water Supply Paper 1827A.

Higashi, T., and K. Wada. 1977. Size fractionation, dissolution analysis, and infrared spectroscopy of humus complexes in Ando soils. *J. Soil Sci.* 28:653–663.

Hoyt, P.B. 1977. Effects of organic matter content on exchangeable Al and pH dependent acidity of very acid soils. *Can. J. Soil Sci.* 57:221–222.

Hoyt, P.B., and R.C. Turner. 1975. Effects of organic materials added to very acid soils on pH, aluminum, exchangeable NH_4 and crop yields. *Soil Sci.* 119:227–237.

Hsu, P.H. 1966. Formation of gibbsite from aging hydroxy-aluminum solutions. *Soil Sci. Soc. Am. Proc.* 30:173–176.

Hsu, P.H. 1967. Effect of salts on the formation of bayerite versus pseudoboehmite. *Soil Sci.* 103:101–110.

Hsu, P.H. 1973. Effect of sulfate on the crystallization of aluminum hydroxide from aging hydroxy-aluminum solutions. In: J. Nicolas (ed.) *Proc. 3d Int. Cong. on Studies of Bauxite and Aluminum Oxides-Hydroxides,* Nice, France, pp. 613–620.

Hsu, P.H. 1975. Precipitation of phosphate from solution using aluminum salts. *Water Res.* 9:1155–1161.

Hsu, P.H. 1977. Aluminum hydroxides and oxyhydroxides. In: J.B. Dixon and S.B. Weed (eds.) *Minerals in Soil Environments.* Soil Science Society of America, Madison, WI, pp. 99–143.

Hsu, P.H. 1984. Aluminum hydroxides and oxyhydroxides in soils: Recent developments. Agronomy Abstracts, 1984 Annual Meetings of the American Society of Agronomy, Crop Science Society of America, and Soil Science Society of America, Las Vegas.

Hsu, P.H., and T.F. Bates. 1964a. Fixation of hydroxy-aluminum polymers by vermiculite. *Soil Sci. Soc. Am. Proc.* 28:763–769.

Hsu, P.H., and T.F. Bates. 1964b. Formation of X-ray amorphous and crystalline aluminum hydroxides. *Miner. Mag.* 33:749–768.

Hsu, P.H., and C.I. Rich. 1960. Aluminum fixation in a synthetic cation exchanger. *Soil Sci. Soc. Am. Proc.* 24:21–25.

Huang, P.M. 1980. Adsorption processes in soil. In: O. Hutzinger, (ed.) *Handbook of Environmental Chemistry.* Springer-Verlag, Berlin, Vol. 2, Part A, pp. 47–59.

Huang, P.M. 1987a. Aluminum and the fate of nutrients and toxic substances in terrestrial and freshwater environments. In: M. Singh (ed.) *Systems and Control Encyclopedia.* Pergamon Press, Oxford, U.K. pp. 262–268.

Huang, P.M. 1987b. Impact of mineralogical research on soil and environmental sciences. In: L.L. Boersma (ed.) *Future Developments of Soil Science.* Golden Anniversary Symposium Special Publication. Soil Science Society of America, Madison, WI, pp. 485–498.

Huang, P.M., and M.L. Jackson. 1965. Mechanism of reaction of neutral fluoride solutions with layer silicates and oxides of soils. *Soil Sci. Soc. Am. Proc.* 29:661–665.

Huang, P.M., and L.M. Kozak. 1970. Adsorption of hydroxy-aluminum polymers by muscovite and biotite. *Nature* 228:1084–1085.

Huang, P.M., and S.Y. Lee. 1969. Effect of drainage on weathering transformations of mineral colloids of some Canadian prairie soils. In: L. Heller (ed.) *Proc. 3d Int. Clay Conf. (Tokyo)* I:541–551.

Huang, P.M., and A. Violante. 1986. Influence of organic acids on crystallization and surface properties of precipitation products of aluminum. In: P.M. Huang and M. Schnitzer (eds.) *Interactions of Soil Minerals with Natural Organics and Microbes.* SSSA Special Publication No. 17. Soil Science Society of America, Madison, WI, pp. 159–221.

Huang, P.M., R. Grover, and R.B. McKercher. 1984. Components and particle size fractions involved in atrazine adsorption by soils. *Soil Sci.* 138:20–24.

Hunt, J.P. 1963. *Metal Ions in Aqueous Solution.* W.A. Benjamin, New York.

Ino, Y., and M. Monsi. 1964. On the decomposition rate of soil organic matter in humic allophane soils at Mount Kirigamine. *Bot. Mag.* (Tokyo) 77:168–175.

Inoue, K., and P.M. Huang. 1984a. Influence of citric acid on natural formation of imogolite. *Nature* 308:58–60.

Inoue, K., and P.M. Huang. 1984b. Effect of humic and fulvic acids on the formation of imogolite. *Agronomy Abstracts*, 1984 Annual Meetings of American Society of Agronomy, Crop Science Society of America, and Soil Science Society of America, Las Vegas, p. 273.

Inoue, K., and P.M. Huang. 1985. Influence of citric acid on the formation of short-range ordered aluminosilicates. *Clays Clay Miner.* 33:312–322.

Inoue, K., and P.M. Huang. 1986. Influence of selected organic ligands on the formation of allophane and imogolite. *Soil Sci. Soc. Am. J.* 50:1623–1633.

Inoue, K., and P.M. Huang. 1987. Effect of humic and fulvic acids on the formation of allophane. In: L.G. Schultz, H. van Olphen, and F.A. Mumpton (eds.) *Proc. 8th Int. Clay Conf., Denver,* pp. 221–226.

Inoue, K., M. Yoshida, and T. Henmi. 1980. The occurrence of allophane in stream-deposits from Shishigahana at the northern foot of Mt. Chokai, Japan. *Clay Sci. (Tokyo)* 5:267–276.

Jackman, R.N. 1964. Accumulation of organic matter under permanent pasture. 2.

Rates of mineralization of organic matter and supply of available nutrients. *N.Z. J. Agric. Res.* 7:472–479.

Jackson, M.L. 1960. Structural role of hydronium in layer silicates during soil genesis. *Trans. 7th Int. Cong. Soil Sci., Madison, WI*, II:445–455.

Jackson, M.L. 1962. Interlayering of expansible layer silicates in soil chemical weathering. *Clays Clay Miner.* 11:29–46.

Jackson, M.L. 1963a. Aluminum bonding in soils: A unifying principle in soil science. *Soil Sci. Soc. Am. Proc.* 27:1–10.

Jackson, M.L. 1963b. Interlayering of expansible layer silicates in soils by chemical weathering. *Clays Clay Miner.* 11:29–46.

Jackson, M.L. 1964. Chemical composition of soils. In: F.E. Bear (ed.) *Chemistry of the Soil*, 2d Ed. Reinhold, New York, pp. 71–141.

Jackson, M.L., S.Y. Lee, J.L. Brown, I.B. Sachs, and J.K. Syers. 1973. Scanning electron microscopy of hydrous oxide crusts intercalated in naturally weathered micaceous vermiculite. *Soil Sci. Soc. Am. Proc.* 73:127–131.

Jander, G., and A. Winkel. 1931. Diffusion coefficients of basic aluminum solutions. *Z. Anorg. Chem.* 200:257–262.

Johansson, G. 1960. On the crystal structures of some basic aluminum salts. *Acta Chem. Scand.* 14:771–773.

Johansson, G. 1962. The crystal structures of $[Al_2(OH)_2(H_2O)_8]$ $(SO_4)_2 \cdot 2H_2O$ and $[Al_2(OH)_2(H_2O)_8]$ $(SeO_4)_2 \cdot 2H_2O$. *Acta Chem. Scand.* 16:403–420.

Johnson, N.M., C.T. Driscoll, J.S. Eaton, G.E. Likens, and W.H. McDowell. 1981. "Acid rain," dissolved aluminum and chemical weathering at the Hubbard Brook Experimental Forest, New Hampshire. *Geochim. Cosmochim. Acta* 45:1421–1437.

Jones, R.C., and G. Uehara. 1973. Amorphous coatings on mineral surfaces. *Soil Sci. Soc. Am. Proc.* 37:792–798.

Julien, A.A. 1879. On the geological action of the humus acids. *Proc. Am. Assoc. Adv. Sci.* 28:311–410.

Kaurichev, I.S., and Y.M. Nozdrunova. 1961. The role of components of water-soluble organic plant residue in forming available (mobile) iron-organic compounds. *Pochvovediniye* 10:1057–1064.

Keller, W.D. 1964. The origin of high alumina clay minerals. A review. *Clays Clay Miner.* 12:129–151.

Kentamma, J. 1955. The hydrolysis of aluminum chloride. *Acad. Sci. Fenn. Ann. Ser. A.*, p. 67.

Kidder, G., and L.W. Reed. 1972. Swelling characteristics of hydroxyaluminum interlayered clays. *Clays Clay Miner.* 20:13–20.

Kinniburgh, D.G., and M.L. Jackson. 1981. Cation adsorption by hydrous metal oxies and clay. In: M.A. Anderson and A.J. Rubin (eds.) *Adsorption of Inorganics at Solid-Liquid Interfaces*. Ann Arbor Science Publishers, Ann Arbor, MI, pp. 91–160.

Kirkland, D.L., and B.F. Hajek. 1972. Formula derivation of Al-interlayered vermiculite in selected soil clays. *Soil Sci.* 114:317–322.

Kissel, D.E., E.P. Gentzsch, and G.W. Thomas. 1971. Hydrolysis of nonexchangeable acidity in soils during salt extractions of exchangeable acidity. *Soil Sci.* 111:293–297.

Kodama, H., and M. Schnitzer. 1980. Effect of fulvic acid on the crystallization of aluminum hydroxides. *Geoderma* 24:195–205.

Kozak, L.M., and P.M. Huang. 1971. Adsorption of hydroxy-Al by certain phyllo-silicates and its relation to K/Ca cation exchange selectivity. *Clays Clay Miner.* 19:95–102.

Kroth, E.M., and J.B. Page. 1947. Aggregration formation in soils with special reference to cementing substances. *Soil Sci. Soc. Am. Proc.* 11:27–34.

Kubota, H. 1956. The hydrolysis of aluminum in dilute solutions, *Diss. Abstr.* 16:864.

Kwong, Ng Kee K.F., and P.M. Huang. 1975. Influence of citric acid on the crystal-lization of aluminum hydroxides. *Clays Clay Miner.* 23:164–165.

Kwong, Ng Kee K.F., and P.M. Huang. 1977. Influence of citric acid on the hydrolytic reactions of aluminum. *Soil Sci. Soc. Am. J.* 41:692–697.

Kwong, Ng Kee K.F., and P.M. Huang. 1979a. The relative influence of low-molecular-weight complexing organic acids on the hydrolysis and precipitation of aluminum. *Soil Sci.* 128:337–342.

Kwong, Ng Kee K.F., and P.M. Huang. 1979b. Nature of hydrolytic precipitation products of aluminum as influenced by low-molecular-weight complexing organic acids. In: M.M. Mortland and V.C. Farmer (eds.) *Proc. 6th Int. Clay Conf., Oxford,* pp. 527–536.

Kwong, Ng Kee K.F., and P.M. Huang. 1979c. Surface reactivity of aluminum hydroxides precipitated in the presence of low-molecular-weight organic acids. *Soil Sci. Soc. Am. J.* 43:1107–1113.

Kwong, Ng Kee K.F., and P.M. Huang. 1981. Comparison of the influence of tannic acid and selected low-molecular-weight organic acids on precipitation products of aluminum. *Geoderma* 26:179–193.

Lind, C.J., and J.D. Hem. 1975. Effects of organic solutes on chemical reactions of aluminum. *U.S. Geol surv. Water Supply Paper 1827-G*, 83 pp.

Lindsay, W.L. 1979. Chemical equilibria in soils, John Wiley, New York.

Lippens, B.C., and J.J. Steggerda. 1970. Active alumina. In: B.G. Linsens (ed.) *Physical and Chemical Aspects of Adsorbents and Catalysts.* Academic Press, New York, pp. 171–211.

Lodding, W. 1961. Gibbsite vermiforms in the Pensouken Formation of New Jersey. *Am. Miner.* 46:394–401.

Luciuk, G.M., and P.M. Huang. 1974. Effect of monosilicic acid on hydrolytic reactions of aluminum. *Soil Sci. Soc. Am. Proc.* 38:235–244.

Lynch, J.M., K.C. Hall, H.A. Anderson, and A. Hepburn. 1980. Organic acids from the anaerobic decomposition of *Agropyron repens* rhizomes. *Phyto-chemistry* 19:1846–1847.

MacEwan, D.M.C. 1950. Some notes on the recording and interpretation of X-ray diagrams of soil clays. *J. Soil Sci.* 1:90–103.

Maksimova, I., V. Mashovets, and V. Yushkevich. 1967. Electrical conductance and structure of sodium aluminates in aqueous solutions. *J. Appl. Chem. USSR* 40:2594–2597.

Marion, S.P., and A.W. Thomas. 1946. Effects of diverse anions on the pH of maximum precipitation of aluminum hydroxide. *J. Colloid Sci.* 1:221–234.

Martin, J.P., and K. Haider. 1986. Influence of mineral colloids on turnover rates of soil organic carbon. In: P.M. Huang and M. Schnitzer (eds.) *Interactions of Soil Minerals with Natural Organics and Microbes.* SSSA Special Publication No. 17. Soil Science Society of America, Madison, WI, pp. 283–304.

Martin, J.P., H. Zunino, P. Peirano, M. Caiozzi, and H. Haider. 1982. Decomposition of ^{14}C-labeled lignins, model humic acid polymers, and fungal melanins in allophanic soils. *Soil Biol. Biochem.* 14:289–293.

Matijevic, E., and B. Tezak. 1953. Detection of polynuclear complex aluminum ions by means of coagulation measurements. *J. Phys. Chem.* 57:951–954.

Matijevic, E., K.G. Mathai, R.H. Otterwill, and M. Kerker. 1961. Detection of metal ion hydrolysis by coagulation. III. Aluminum. *J. Phys. Chem.* 65:826–830.

May, H.M., P.A. Helmke, and M.L. Jackson. 1979. Gibbsite solubility and thermodynamic properties of hydroxy-aluminum ions in aqueous solution at 25°C. *Geochim. Cosmochim. Acta* 43:861–865.

McKeague, J.A., and H. Kodama. 1981. Imogolite in cemented horizons of some British Columbia soils. *Geoderma* 25:189–197.

McKeague, J.A., M.V. Cheshire, F. Andreux, and J. Berthelin. 1986. Organomineral complexes in relation to pedogenesis. In: P.M. Huang and M. Schnitzer (eds.) *Interactions of Soil Minerals with Natural Organics and Microbes.* SSSA Speical Publication No. 17. Soil Science Society of America, Madison, WI, pp. 549–592.

McLean, E.O. 1976. Chemistry of soil aluminum. *Comm. Soil Sci. Plant Anal.* 7:619–636.

McLean, E.O., D.C. Reicosky, and C. Lakshmanan. 1965. Aluminum in soils. VII. Interrelationships of organic matter, liming and exchangeable aluminum with "permanent charge" (KCI) and pH-dependent cation exchange capacity of surface soils. *Soil Sci. Soc. Am. Proc.* 29:374–378.

Mesmer, R.E., and C.F. Baes Jr. 1971. Acidity measurements at elevated temperatures. V. Aluminum ion hydrolysis. *Inorg. Chem.* 10:2290–2296.

Milton, C., E.J. Dwornik, and R.B. Finkelman. 1975. Nordstrandite, $Al(OH)_3$, from the Green River formation in Rio Blanco County, Colorado. *Am. Miner.* 60:285–291.

Mitchell, B.D., V.C. Farmer, and W.J. McHardy. 1964. Amorphous inorganic materials in soils. *Adv. Agron.* 16:327–383.

Moghimi, A., M.E. Tate, and J.M. Oades. 1978. Characterizations of rhizosphere products, especially 2-ketogluconic acid. *Soil Biol. Biochem.* 10:283–287.

Moolenaar, R.J., J.C. Evans, and L.D. McKeener. 1970. The structure of the aluminate ion in solution at high pH. *J. Phys. Chem.* 74:3629–3636.

Murmann, R.K. 1964. *Inorganic Complex Compounds.* Reinhold, New York.

Nail, S.L., J.L. White, and S.L. Hem. 1976a. Structure of aluminum hydroxide gel. 1. Initial precipitate. *J. Pharm. Sci.* 65:1188–1191.

Nail, S.L., J.L. White, and S.L. Hem. 1976b. Structure of aluminum hydroxide gel. 2. Aging mechanism. *J. Pharm. Sci.* 65:1192–1195.

Nordstrom, D.K. 1982. The effect of sulfate on aluminum concentrations in natural water: Some stability relations in the system Al_2O_3-SO_3-H_2O at 298K. *Geochim. Cosmochim. Acta* 46:681–692.

Novak, R.J., H.L. Motto, and L.A. Douglas. 1971. The effect of time and particle size on mineral alteration in severeal Quaternary soils in New Jersey and Pennsylvania, U.S.A. In: D.H. Yaalon (ed.) *Paleopedology—Origin, Nature, and Dating of Paleosols.* Israel University Press, Jerusalem, pp. 211–224.

Oades, J.M. 1978. Mucilages at the root surface. *J. Soil Sci.* 29:1–16.

Okura, T., K. Goto, and T. Votuyanagi. 1962. Forms of aluminum determined by an 8-quinolinolate extraction method. *Anal. Chem.* 34:581–582.

Parfitt, R.L. 1978. Anion adsorption by soils and soil materials. *Adv. Agron.* 30:1–50.

Parfitt, R.L., and T. Henmi. 1980. Structure of some allophanes from New Zealand. *Clays Clay Miner.* 28:285–294.

Parthasarthy, N., and J. Buffle. 1985. Study of polymeric aluminum (III) hydroxide solutions for application in waste water treatment. Properties of the polymer and optional conditions of preparation. *Water Res.* 19:25–36.

Patterson, J.H., and S.Y. Tyree. 1973. A light scattering study of the hydrolytic polymerization of aluminum. *J. Colloid Interface Sci.* 43:384–398.

Paul, E.A., and P.M. Huang. 1980. Chemical aspect of soil. In: O. Hutzinger (ed.) *Handbook of Environmental Chemistry*, Vol. 1, Part A, Springer-Verlag, New York, pp. 69–86.

Pearson, R.W., and L.E. Ensminger. 1949. Types of clay minerals in Alabama soils. *Soil Sci. Soc. Am. Proc.* 13:153–156.

Perrott, K.W. 1981. The nature of cationic aluminum species on the cation exchange surface of mica. *J. Colloid Interface Sci.* 82:136–140.

Pionke, H.B., and R.B. Corey. 1967. Relations between acidic aluminum and soil pH, clay and organic matter. *Soil Sci. Soc. Am. Proc.* 31:749–752.

Ragland, J.L., and N.T. Coleman. 1960. The hydrolysis of aluminum salts in clay and soil systems. *Soil Sci. Soc. Am. Proc.* 24:457–460.

Rao, D.N., and D.S. Mikkelsen. 1977. Effect of rice straw additions on production of organic acids in a flooded soil. *Plant Soil* 47:303–311.

Rausch, M.V., and H.D. Bale. 1964. Small-angle X-ray scattering from hydrolyzed Al nitrate solutions. *J. Chem. Phys.* 40:3391–3395.

Rich, C.I. 1968. Hydroxy interlayers in expansible phyllosilicates. *Clays Clay Miner.* 16:15–30.

Rich, C.I., and S.S. Obenshain. 1955. Chemical and clay mineral properties of a Red-Yellow Podzolic soil derived from Muscovite schist. *Soil Sci. Soc. Am. Proc.* 19:334–339.

Rich, C.I., and W.R. Black. 1964. Potassium exchange as affected by cation size, pH, and mineral structure. *Soil Sci.* 97:384–390.

Richburg, J.S., and F. Adams. 1970. Solubility and hydrolysis of aluminum in soil solution and saturated-paste extracts. *Soil Sci. Soc. Am. Proc.* 34:728–734.

Robert, M., and J. Berthelin. 1986. Role of biological and biochemical factors in soil mineral weathering. In: P.M. Huang and M. Schnitzer (eds.) *Interactions of Soil Minerals with Natural Organics and Microbes*. SSSA Special Publication No. 17. Soil Science Society of America, Madison, WI, pp. 453–495.

Roberson, C.E., and J.D. Hem. 1969. Solubility of aluminum in the presence of hydroxide, fluoride, and sulfate. *U.S. Geol. Surv. Water Supply Paper 1827C*, 37 pp.

Ross, D.J. 1983. Invertase and amylase activities as influenced by clay minerals, soil clay fractions and topsoils under grassland. *Soil Biol. Biochem.* 15:287–293.

Ross, G.J., and H. Kodama. 1979. Evidence for imogolite in Canadian soils. *Clays Clay Miner.* 27:297–300.

Ross, G.J., and R.C. Turner. 1971. Effect of different anions on the crystallization of aluminum hydroxide in partially neutralized aqueous aluminum salt systems. *Soil Sci. Soc. Am. Proc.* 35:389–392.

Rovira, A.D., and B.M. McDougall. 1967. Microbiology and biochemical aspects of the rhizosphere. In: A.D. McLaren and G.H. Peterson (eds.) *Soil Biochemistry*, Vol. 1. Marcel Dekker, New York, pp. 417–463.

Saini, G.R., A.A. MacLean, and J.J. Doyle. 1966. The influence of some physical and chemical properties on soil aggregation and response to VAMA. *Can. J. Soil Sci.* 46:155–160.

Sato, K., and I. Yamane. 1967. Effect of temperature on the decomposition of glucose in a flooded soil. *J. Soil Sci. Manure Jpn.* 37:547–551.

Sawhney, B.L. 1968. Aluminum interlayers in layer silicates. Effect of OH/Al ratio of Al solution, time of reaction, and type of structure. *Clays Clay Miner.* 16:157–163.

Schnitzer, M. 1968. Reactions between organic matter and inorganic constituents. *Trans. 9th Int. Cong. Soil Sci., Adelaide, Australia,* I:635–644.

Schnitzer, M. 1977. Recent findings on the characterization of humic substances extracted from soils from widely differing climataic zones. In: *Soil Organic Matter Studies*, Vol. II. IAEA, Vienna, SM-211/7.

Schnitzer, M. 1986. Binding of humic substances by soil mineral colloids. In: P.M. Huang and M. Schnitzer (eds.) *Interactions of Soil Minerals with Natural Organics and Microbes*. SSSA Special Publication No. 17. Soil Science Society of America, Madison, WI, pp. 77–101.

Schnitzer, M., and H. Kodama. 1977. Reactions of minerals with soil humic substances. In: J.B. Dixon and S.B. Weed (eds.) *Minerals in Soil Environments*. Soil Science Society of America, Madison, WI, pp. 741–770.

Schnitzer, M., and S.I.M. Skinner. 1965. Organo-metallic interactions in soils. 4. Carboxyl and phenolic hydroxyl groups in organic matter and metal retention. *Soil Sci.* 99:278–284.

Schoen, R., and E.C. Roberson. 1970. Structures of aluminum hydroxide and geochemical implication. *Am. Miner.* 55:43–77.

Schofield, R.K., and A.W. Taylor. 1954. Hydrolysis of aluminum salt solutions. *J. Chem. Soc.* 4445–4448.

Schwertmann, U., and M.L. Jackson. 1963. Hydrogen-aluminum clays: A third buffer range appearing in potentiometric titration. *Science* 139:1052–1053.

Schwertmann, U., and M.L. Jackson. 1964. Influence of hydroxy aluminum ions on pH titration curves of hydronium-aluminum clays. *Soil Sci. Soc. Am. Proc.* 28:179–183.

Schwertmann, U., and R.M. Taylor. 1982. The significance of oxides for the surface properties of soils and the usefulness of synthetic oxides as models for their study. *Bull. Int. Soc. Soil Sci.* 60:62–66.

Serna, C.J., J.L. White, and S.L. Hem. 1977. Anion-aluminum hydroxide gel interactions. *Soil Sci. Soc. Am. J.* 41:1009–1013

Shoemaker, H.E., E.O. McLean, and P.F. Pratt. 1961. Buffer methods for determining lime requirements of soils with appreciable amounts of extractable aluminum. *Soil Sci. Soc. Am. Proc.* 25:274–277.

Shoji, S., and J. Masui. 1972. Amorphous clay minerals of recent volcanic ash soils. 3. Mineral composition of fine clay fractions. *J. Sci. Soil Manure Jpn.* 43:187–193.

Sillen, L.G. 1959. Quantitative studies of hydrolytic equilibria. *Q. Rev.* 13:146–168.

Sillen, L.G. 1961. On equilibria in systems with polynuclear complex formation. *Chem. Scand.* 15:1981–1986.

Singer, A., and P.M. Huang. 1986. Effect of humic acid on Al-interlayering in montmorillonite. Vol. IV, XIII Congr. Int. Soc. Soil Sci., Hamburg, abstract, pp. 1478–1479.

Singh, S.S. 1969. Neutralization of dilute aqueous aluminium sulphate solutions with a base. *Can. J. Chem.* 47:663–667.

Singh, S.S. 1972. The effect of temperature on the ion activity product (Al) (OH)^3 and its relation to lime potential and degree of base saturation. *Soil Sci. Soc. Am. Proc.* 36:47–50.

Singh, S.S., and J.E. Brydon. 1967. Precipitation of aluminum by calcium hydroxide in the presence of Wyoming bentonite and sulfate ions. *Soil Sci.* 103:162–167.

Smith, R.W. 1971. Reactions among equilibrium and nonequilibrium aqueous species of aluminum hydroxy complexes. *Adv. Chem. Ser. Am. Chem. Soc.* 106:250–270.

Smith, R.W., and J.D. Hem. 1972. Effect of aging on aluminum hydroxide complexes in dilute aqueous solutions. *U.S. Geol Surv. Water Supply Paper 1827-D.*

Soil Survey Staff. 1975. Soil taxonomy: A basic system of soil classification for making and interpreting soil surveys. *Agric. Handb. No. 436.* U.S. Government Printing Office, Washington.

Souza Santos, P., A. Vallejo-Freire, and H.L. Souza Santos. 1953. Electron microscope studies on the aging of amorphous colloid aluminum hydroxide. *Kolloid Z.* 133:101–107.

Sprengel, C. 1826. Uber Pflanzenhumus, Humussäure humassaure Salze. *Kastners Arch. Naturlehre* 8:145–220.

Stevenson, F.J. 1967. Organic acid in soil. In: A.D. McLaren and G.H. Peterson (eds.) *Soil Biochemistry*, Vol. I. Marcel Dekker, New York, pp. 119–146.

Stevenson, F.J. 1982. *Humus Chemistry.* John Wiley, New York.

Stevenson, F.J., and M.S. Ardakani. 1972. Organic matter reactions involving micronutrients in soils. In: J.J. Mortvedt, P.M. Giordano, and W.L. Lindsay (eds.) *Micronutrients in Agriculture.* Soil Science Society of America, Madison, WI, pp. 79–114.

Stol, R.J., A.D. van Helden, and P.L. de Bruyn. 1976. Hydrolysis-precipitation studies of aluminum solution. II. A kinetic study and a model. *J. Colloid Interface Sci.* 57:115–131.

Stotzky, G. 1986. Influence of soil mineral colloids on metabolic processes, growth, adhesion, and ecology of microbes and viruses. In: P.M. Huang and M. Schnitzer (eds.) *Interactions of Soil Minerals with Natural Organics and Microbes.* SSSA Special Publication No. 17. Soil Science Society of America, Madison, WI, pp. 305–428.

Tait, J.M., N. Yoshinaga, and B.D. Mitchell. 1978. The occurrence of imogolite in some Scottish soils. *Soil Sci. Plant Nutr. (Tokyo)* 24:145–151.

Tamura, T. 1957. Identification of the 14Å clay mineral component. *Am. Miner.* 42:107–110.

Taylor, A.W., and E.L. Gurney. 1965. Precipitation of phosphate by iron oxide and aluminum hydroxide from solutions containing calcium and potassium. *Soil Sci. Soc. Am. Proc.* 29:18–22.

Teagarden, D.L., J.F. Kozlowski, J.L. White, and S.L. Hem. 1981. Aluminum chlorohydrate. I. Structure Studies. *J. Pharm. Sci.* 70:758–761.

Teagarden, D.L., S.L. Hem, and J.L. White. 1982. Conversion of aluminum chlorohydrate to aluminum hydroxide. *J. Soc. Cosmet. Chem.* 33:281–295.

Tettenhorst, R., and A. Hoffmann. 1980. Crystal chemistry of boehmite. *Clays Clay Miner.* 28:373–380.

Thenard, P. 1870. Observations sur le mémoire de M. Friedel. *C.R.* 70:1412–1414.

Thomas, G.W. 1975. The relationship between organic matter content and exchangeable aluminum in acid soil. *Soil Sci. Soc. Am. Proc.* 39:591.

Thomas, G.W. 1977. Historical developments in soil chemistry: Ion exchange. *Soil Sci. Soc. Am. J.* 41:230–238.

Thomas, G.W., and W.L. Hargrove. 1984. The chemistry of soil acidity. In: F. Adam (ed.) *Soil Acidity and liming. Agronomy* No. 12. American Society of Agronomy, Crop Science Society of America, and Soil Science Society of America, Madison, WI, pp. 3–56.

Tokashiki, Y., and K. Wada. 1975. Weathering implications of the mineralogy of clay fractions of two Ando soils, Kyushu. *Geoderma* 14:47–62.

Toy, A.D., T.D. Smith, and J.R. Pilbrow, 1973. Aluminum-27 nuclear magnetic resonance in aqueous solutions of its chelates with hydroxy carboxylic acids. *Aust. J. Chem.* 26:1889–1892.

Tsai, P.P., and P.H. Hsu. 1984. Studies of aged OH-Al solutions using kinetics of Al-ferron reactions and sulfate precipitation. *Soil Sci. Soc. Am. J.* 48:59–65.

Tsai, P.P., and P.H. Hsu. 1985. Aging of partially neutralized aluminum solutions of sodium hydroxide/aluminum molar ratio = 2.2. *Soil Sci. Soc. Am. J.* 49:1060–1065.

Turner, R.C. 1965. Some properties of aluminum hydroxide precipitated in the presence of clays. *Can. J. Soil Sci.* 45:331–336.

Turner, R.C. 1969. Three forms of aluminum in aqueous systems determined by 8-quinolinolate extraction methods. *Can. J. Chem.* 47:2521–2527.

Turner, R.C. 1971. Kinetics of reactions of 8-quinolinol and acetate with hydroxyaluminum species in aqueous solutions. 2. Initial solid phases. *Can. J. Chem.* 49:1688–1690.

Turner, R.C. 1976. Effect of aging on properties of polynuclear hydroxyaluminum cations. *Can. J. Chem.* 54:1528–1534.

Turner, R.C., and J.E. Brydon. 1965. Factors affecting the solubility of $Al(OH)_3$ precipitated in the presence of montmorillonite. *Soil Sci.* 100:176–181.

Turner, R.C., and J.E. Brydon. 1967a. Effect of length of time of reaction on some properties of suspension of Arizona bentonite, illite, and kaolinite in which aluminum hydroxide is precipitated. *Soil Sci.* 103:111–117.

Turner, R.C., and J.E. Brydon. 1967b. Removal of interlayer aluminum hydroxide from montmorillonite by seeding the suspension with gibbsite. *Soil Sci.* 104:332–335.

Turner, R.C., and G.J. Ross. 1970. Conditions in solution during the formation of gibbsite in dilute Al salt solutions. IV. Effect of Cl^- concentration and temperature and a proposed mechanism for gibbsite formation. *Can. J. Chem.* 48:723–729.

Turner, R.C., and W. Sulaiman. 1971. Kinetics of reactions of 8-quinolinol and acetate with hydroxyaluminum species in aqueous solutions. I. Polynuclear hydroxy-aluminum cations. *Can. J. Chem.* 49:1683–1687.

Underwood, E.J. 1977. *Trace Elements in Human and Animal Nutrition*, 4th Ed. Academic Press, New York.

Vaughan, D.E.W., and R. Lussier. 1980. Preparation of molecular sieves based on pillared interlayered clays (PILC). In: L.V. Rees (ed.) *Proc. 5th Int. Conf. Zeolites.* Heyden, London.

Vedder, W., and D.A. Vermilyea. 1969. Aluminum + water reaction. *Trans. Faraday Soc.* 65:561–584.

Vermeulen, A.C., J.W. Geus, R.J. Stol, and P.L. DeBruyn. 1975. Hydrolysis-precipitation studies of aluminum (III) solutions. I. Titration of acidified aluminum nitrate solutions. *J. Colloid Interface Sci.* 51:449–458.

Violante, A., and P.M. Huang. 1984. Characteristics and surface properties of pseudoboehmites formed in the presence of selected organic and inorganic ligands. *Soil Sci. Soc. Am. J.* 48:1193–1201.

Violante, A., and P.M. Huang. 1985. Influence of inorganic and organic ligands on the formation of aluminum hydroxides and oxyhydroxides. *Clays Clay Miner.* 33:181–192.

Violante, A., and M.L. Jackson. 1979. Crystallization of nordstrandite in citrate systems in the presence of montmorillonite. In: M.M. Mortland and V.C. Farmer (eds.) *Proc. 6th Int. Clay Conf. (Oxford)* 1:517–525.

Violante, A., and M.L. Jackson. 1981. Clay influence on the crystalization of aluminum hydroxide polymorphs in the presence of citrate, sulfate, or chloride. *Geoderma* 25:199–214.

Violante, A., and P. Violante. 1978. Influence of carboxylic acids on the stability of chlorite-like complexes and on the crystallization of Al(OH)$_3$ polymorphs. *Agrochimica* 22:335–343.

Violante, A., and P. Violante. 1980. Influence of pH, concentration and chelating power of organic anions on the synthesis of aluminum hydroxides and oxyhydroxides. *Clays Clay Miner.* 28:425–434.

Wada, K. 1977. Allophane and imogolite. In: J.B. Dixon and S.B. Weed (eds.) *Minerals in Soil Environments.* Soil Science Society of America, Madison, WI, pp. 603–638.

Wada, K. 1979. Structural formulas of allophanes. In: M.M. Mortland and V.C. Farmer (eds.) *Proc. 6th Int. Clay Conf., Oxford*, pp. 537–545.

Wada, K. 1980. Mineralogical characteristics of Andisols. In: B.K.G. Theng (ed.) *Soils with Variable Charge.* New Zealand Society for Soil Science, Palmerston North, New Zealand, pp. 87–107.

Wada, K. 1981. Amorphous clay minerals—chemical composition, crystalline state, synthesis, and surface properties. In: H. van Olphen and F. Veniale (eds.) *Proc. 7th Int. Clay Conf., Bologna, Pavia*, pp. 385–398.

Wada, K., and D.J. Greenland. 1970. Selective dissolution and differential infrared spectroscopy for characterization of "amorphous" constituents in soil clays. *Clay Miner.* 8:241–254.

Wada, K., and M.E. Harward. 1974. Amorphous clay constituent of soils. *Adv. Agron.* 26:211–260.

Wada, K., and T. Higashi. 1976. The categories of aluminum- and iron-humus complexes in Ando soils determined by selective dissolution. *J. Soil. Sci.* 27:357–368.

Wada, S.I., A. Eto, and K. Wada. 1979. Synthetic allophane and imogolite. *J. Soil. Sci.* 30:347–355.

Wall, J.R.D., E.B. Wolfenden, E.H. Beard, and T. Deans. 1962. Nordstrandite in soil from West Sarawak, Borneo. *Nature* 196:264–265.

Walters, E.H. 1916. Presence and origin of volatile fatty acids in soils. *Science* 44:217.

Wang, M.C., and P.M. Huang. 1986. Polyphenol transformations as catalyzed by oxides of Mn, Fe, Al and Si. *Agronomy Abstracts*, pp. 173–174. *1986 Annual*

Meetings of the American Society of Agronomy, Crop Science Society of America, and Soil Science Society of America, New Orleans.

Wang, M.C., and P.M. Huang. 1987. Catalytic polymerization of hydroquinone by nontronite. *Can. J. Soil Sci.* 67:867–875.

Wang, T.S.C., P.M. Huang, C.-H. Chou, and J.-H. Chen. 1986. The role of soil minerals in the abiotic polymerization of phenolic compounds and formation of humic substances. In: P.M. Huang and M. Schnitzer (eds.) *Interactions of Soil Minerals with Natural Organics and Microbes.* SSSA Special Publication No. 17. Soil Science Society of America, Madison, WI, pp. 251–281.

Wang, T.S.C., M.-M. Kao, and S.W. Li. 1978. A new proposed mechanism of formation of soil humic substances. In: *Studies and Essays in Commemoration of the Golden Jubilee of Academia Sinica.* Academia Sinica, Taipei, pp. 357–372.

Wang, T.S.C., T.K. Yang, and T.T. Chuang. 1967. Soil phenolic acids as plant growth inhibitors. *Soil Sci.* 103:239–246.

Weast, R.C. 1978. *CRC Handbook of Chemistry and Physics*, 58th Ed. CRC Press, Boca Raton, FL.

Weldon, T.A., and J.C. Hide. 1942. Some chemical properties of soil organic matter and of sesquioxides assoicated with aggregation in soils. *Soil Sci.* 54:343–352.

Wells, A.F. 1962. *Structural Inorganic Chemistry.* 3d Ed. Oxford University Press, London, 552 pp.

Wells, N., C.W. Childs, and C.J. Downes. 1977. Silica Springs, Tongariro National Park, New Zealand—anaylses of the spring water and characterization of the aluminosilicate deposit. *Geochim. Cosmochim. Acta* 41:1497–1506.

White, J.L., and S.L. Hem. 1975. Role of carbonate in aluminum hydroxide gel established by Raman and IR anaylsis. *J. Pharm. Sci.* 64:468–469.

Whitehead, D.C. 1964. Identification of *p*-hydroxybenzoic, vanillic, *p*-coumaric, and ferulic acids in soils. *Nature* 202:417–418.

Whitehead, D.C., H. Dibb, and R.D. Hartley. 1981. Extractant pH and the release of phenolic compounds from soils, plant roots and leaf litter. *Soil Biol. Biochem.* 13:343–348.

Yase, Y. 1980. The role of aluminum in CNS degeneration with interaction of calcium. In: L. Liss (ed.) *Aluminum Neurotoxicity.* Pathotox Publishers, Park Forest South, IL, pp. 101–109.

Yoldas, B.E. 1973. Hydrolysis of aluminum alkoxides and bayerite conversion. *J. Appl. Chem. Biotechnol.* 23:803–809.

Zunino, H., and J.P. Martin. 1977. Metal-binding organic macromolecules in soil. I. Hypothesis interpreting the role of soil organic matter in the translocation of metal ions from rocks to biological system. *Soil Sci.* 123:65–76.

Zunino, H., F. Borie, S. Aguilera, J.P. Martin, and K. Haider. 1982. Decomposition of ^{14}C-labeled glucose, plant and microbial products, and phenols in volcanic ash–derived soils of Chile. *Soil Biol. Biochem.* 14:37–43.

Improving and Sustaining Productivity in Dryland Regions of Developing Countries[*]

J.L. Steiner, J.C. Day, R.I. Papendick,
R.E. Meyer, and A.R. Bertrand

I. Introduction

Interest in dryland and rain-fed farming systems has increased significantly in recent years in many regions of the world because of rapidly increasing human populations coupled with low productivity gains, escalating water development costs for new irrigation projects, and high operation and maintenance costs associated with irrigated agriculture.

---

[*]Contribution from USDA, Agricultural Research Service, Bushland, Texas, and Pullman, Washington; USDA, Economic Research Service; and U.S. Agency for International Development. This chapter was prepared by U.S. Government employees as part of their official duties. However, the opinions expressed are those of the authors and do not imply official policies of the U.S. Department of Agriculture. Copyright is not claimed.

The terms "rain-fed" and "dryland" are often used synonymously, but in fact, they refer to vastly different physical and biological systems. Both rain-fed and dryland systems exclude irrigation, but beyond that, they can differ significantly. Stewart and Burnett (1987) characterized dryland agricultural systems as those that emphasize water conservation, sustainable crop yields, limited inputs for soil fertility maintenance, and wind and water erosion constraints, whereas rain-fed systems in more humid zones often emphasize disposal of excess water, maximum crop yields, and substantial inputs of fertilizer. Oram (1980) also distinguished between rain-fed farming and dryland agriculture stating that dryland agriculture—as opposed to rain-fed farming—is defined as husbandry under conditions of moderate to severe moisture stress during a substantial portion of the year, which require special cultural techniques and adapted crops and systems for successful and stable agricultural production. Such conditions generally occur in regions classified as semiarid or arid. Pastoral systems are an important part of dryland agriculture, because in some areas, particularly arid areas, they constitute the sole form of agricultural use. However, the purpose of this paper is to discuss dryland crop production in developing areas, including a description of dryland areas, major constraints to dryland productivity, and measures required to alleviate these constraints.

A. Definition of Dryland Areas

Although dryland regions can be described in general terms, specific delineations of dryland areas are difficult. Two contrasting and influential definitions are those based on an aridity index and on length of the growing period.

1. Aridity Index

The United Nations Conference on Desertification (UNESCO, 1977) defined bioclimatic zones based on the climatic aridity index: P/ETP, where P = precipitation and ETP = potential evapotranspiration calculated by the method of Penman (Doorenbos and Pruitt, 1977), taking into account atmospheric humidity, wind, and solar radiation. The zones established by the conference (UNESCO, 1977) were as follows: (1) the *hyperarid zone* (P/ETP < 0.03), consisting of areas largely void of vegetation except for ephemerals and shrubs in river beds and which are virtually unsettled; (2) the *arid zone* (0.03 < P/ETP < 0.20), comprising dryland areas with sparse perennial and annual vegetation utilized mainly by pastoral systems; (3) the *semiarid zone* (0.20 < P/ETP < 0.50), including steppe or tropical shrubland with a discontinuous herbaceous layer and increased frequency of perennials where dryland farming is widely practiced; and (4) the *subhumid zone* (0.50 < P/ETP < 0.75), characterized by more dense vegetation where rain-fed farming is widely practiced with crops adapted to seasonal drought.

2. Growing Period

The Food and Agriculture Organization (FAO) of the United Nations (FAO, 1978a) used the growing period as the basis for assessing climatic resources in developing countries. The growing period is the number of days during a year when precipitation exceeds half the potential evapotranspiration, plus a period required to use an assumed 100 mm of water from excess precipitation (or less, if not available) stored in the soil profile. A normal growing period by their classification must exhibit a humid period, having an excess of precipitation over potential evapotranspiration. Growing periods that did not include a humid period were classified as intermediate periods. Finally, areas where precipitation never exceeded half the potential evapotranspiration were classified as dry with no growing period. Additionally, any time interval during the period when water is available is excluded if the temperature is too low for crop growth (mean temperature below 6.5°C). Calculation of the growing period is based on a simple water balance model, comparing precipitation (P) with potential evapotranspiration (ETP).

Areas having a growing period between 1 and 74 days are classified as arid, and areas with growing periods between 75 and 119 are considered semiarid. The regions of Africa (FAO, 1978a), southwest Asia (FAO, 1978b), southeast Asia (FAO, 1980), and South and Central America (FAO, 1981) have been characterized by this classification system.

3. Example Locations

The average monthly precipitation, potential evapotranspiration, and half-potential evapotranspiration for three locations are presented in Figure 1. All three locations are classified as semiarid by the aridity index ($0.2 <$ P/ETP < 0.5). However, by the FAO growing period classification, only the Rajkot, India, location is classified as semiarid. This location has a growing period of 96 days, which is within the 75-to-119-day range classified as semiarid. The Amman, Jordan, location has a growing period (P > 0.5 ETP) in excess of 120 days, so this location would not be considered semiarid. Bushland, Texas, located in a major dryland farming region in the United States, would be classified as dry, with a 0-day growing period, because the average monthly precipitation never exceeds 0.5 ETP.

These examples point out that there is still a lot of imprecision in defining and classifying climatic zones. Each classification scheme presented, as well as many others in the literature, has advantages for specific purposes and locations, but subjective judgment is required for their interpretation.

Although it is difficult to clearly delineate dryland areas, they do have certain common climatic characteristics. Oram (1980) lists these as (1) low total rainfall with at least one pronounced dry season (and sometimes two) so that lack of moisture puts a ceiling on year-round cropping even though

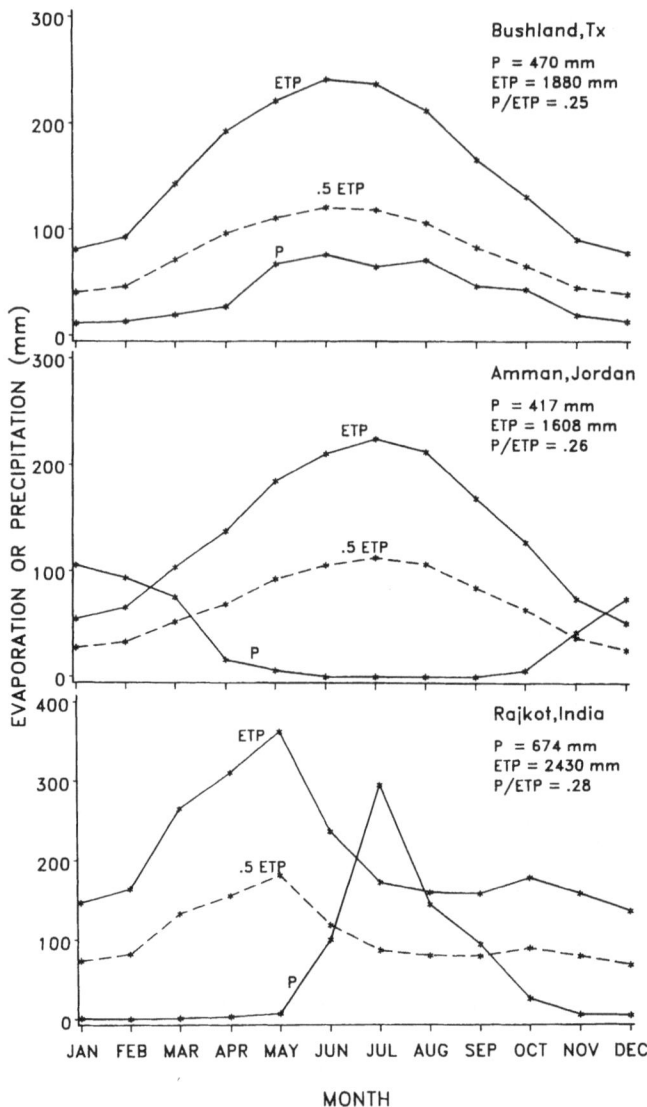

Figure 1. Agroclimatic characteristics of three representative dryland locations.

it may be adequate for one crop; (2) highly variable and unreliable precipitation during the rainy season, with large year-to-year differences in total rainfall and its distribution, and from month to month within seasons; (3) increasing unreliability and variability with decreasing annual rainfall; (4) potential evapotranspiration exceeding precipitation for at least 7 months of the year; and (5) very high-intensity rainstorms leading to high runoff and erosion.

B. Geographic Distribution

The dryland agricultural regions of the world, as characterized by Dregne (1982), are shown in Figure 2. Dregne included areas having a growing period of 90–270 days. These areas are essentially those commonly referred to as semiarid and subhumid. In the wetter parts of these areas, moisture is not a limiting factor for most of the year. It is during the few months of the dry season that moisture deficiencies restrict crop growth even though temperatures are favorable.

A more conservative, and perhaps a more accurate, representation of dryland areas is presented in Figure 3 for the developing countries of the world. This map is based on the FAO (1978a) growing period classification described earlier and represents arid and semiarid regions. The arid regions are used mostly for pastoral systems, and crop production is largely restricted to the semiarid regions.

The extent of the dryland areas can be more clearly shown by the data presented in Table 1. Very large portions of Africa and southwest Asia are in the low-rainfall regions. The 1975 human population numbers for the respective areas are shown in Table 2. Southwest Asia has about 30% of its total population in the dryland regions, and Africa and southeast Asia have larger absolute numbers of people in the dryland areas than southwest Asia, although they have lower percentages of their population in dryland areas. The population numbers reported are 1975 figures and would be substantially higher now, because the rate of population growth in many of the dryland regions is in excess of 2.5% annually (World Bank, 1986).

C. Precipitation Patterns

Four kinds of rainfall distribution patterns occur in dryland regions—winter, summer, continental, and multimodal (Dregne, 1982). The winter rainfall type, also known as Mediterranean, is characterized by most of the precipitation occurring in 8 or 9 cool months, with the other months completely or nearly dry. This pattern is found along the west coast of South and North America, in southern and western Australia, and around the Mediterranean Sea as far east as Iran. The data for Amman, Jordan, in Figure 1 represents this pattern. The summer rainfall type has precipitation concentrated in 3–5 summer months, with the remainder of the year being dry. Rainfall in the Sahelian countries south of the Sahara follows this pattern, with the rainy season becoming longer with increasing distance south from the Sahara. Botswana, northeastern Brazil, and northern Australia have similar patterns. The pattern shown in Figure 1 for Rajkot, India, is fairly typical of this summer rainfall type. "Continental" rainfall patterns have precipitation distributed throughout the year, with pronounced summer peaks and with lesser amounts during the winter, often occurring as snow. This pattern is dominant in regions of central North America, the Soviet Union, China, and Argentina and is represented in

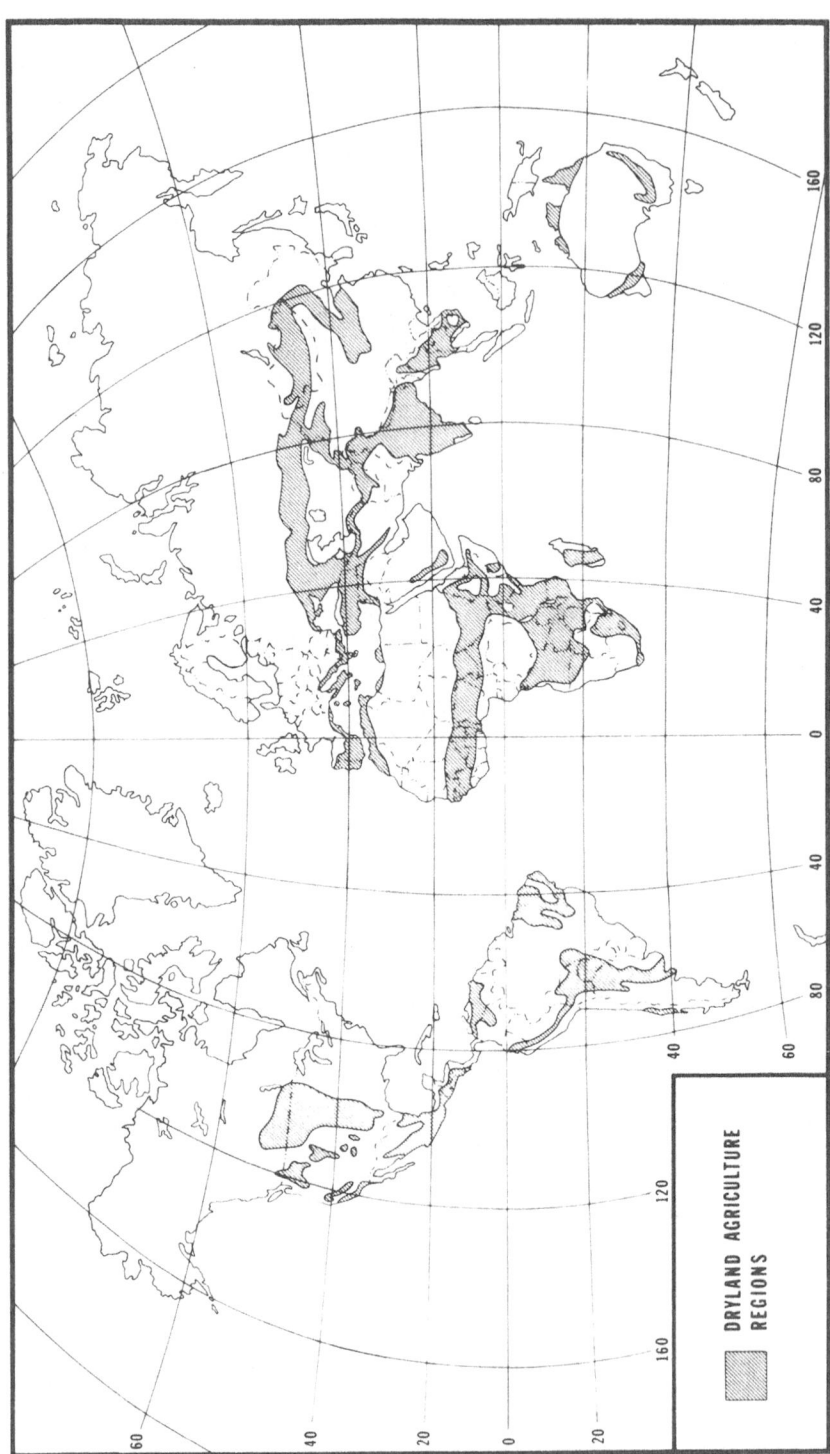

Figure 2. Dryland areas of the world (Dregne, 1982).

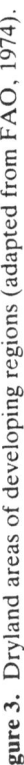

Figure 3. Dryland areas of developing regions (adapted from FAO, 1974).

Table 1. Land area of different growing season zones in developing countries within major geographic regions (in millions of hectares)

Growing days (number)	Africa[a]	Southwest Asia[a]	Southeast Asia[a]	Central America[a]	South America[a]	East Asia[b]
0—Cold	9.1	113.7	47.7	0.8	60.8	NA
0—Dry	846.7	369.7	39.2	35.9	81.2	NA
1-74[c]	*487.9*	*72.6*	*54.6*	*62.2*	*114.6*	*27.7*
75-119	*230.7*	*61.9*	*55.0*	*12.1*	*116.7*	*70.4*
120–179	314.7	37.0	146.9	50.7	113.7	NA
180–269	548.1	20.2	249.7	66.9	293.6	NA
270–365	440.9	2.3	304.5	43.0	989.6	NA
Total	2878.1	677.4	897.6	271.6	1770.2	954.6

Consolidated from data in [a]FAO, 1982; [b]FAO, 1987.

[c]Figures in italics represent arid and semiarid regions.

Table 2. Human population in millions in 1975 in developing countries within major geographic regions living in different growing season zones[a]

Growing days (number)	Africa	Southwest Asia	Southeast Asia	Central America	South America	East Asia
0—Cold	3.7	36.3	11.5	4.2	8.0	NA
0—Dry	48.8	20.2	27.4	3.0	8.1	NA
1-74[b]	*31.0*	*16.1*	*46.2*	*8.5*	*9.4*	*NA*
75-119	*32.7*	*24.4*	*101.1*	*3.8*	*21.1*	*NA*
120–179	98.3	17.9	286.0	25.5	25.2	NA
180–269	113.6	19.6	385.5	45.2	41.0	NA
270–365	78.8	1.8	260.0	16.4	103.0	NA
Total	406.9	136.3	1117.7	106.6	215.8	NA

[a]Consolidated from data in FAO, 1982.

[b]Figures in italics represent arid and semiarid regions.

Figure 1 by the data for Bushland, Texas. The multimodal type, with two or more short rainy periods interspersed with dry months, occurs in parts of eastern Africa and southeast Asia.

D. Present Situation

Land degradation is common in many dryland regions and can lead to desertification. Degradation processes can be arrested and even reversed, but desertification is a continuously degrading process going through several stages before reaching an irreversible stage. Desertification is the impoverishment of terrestrial ecosystems under the impact of human pressure. It results in deterioration of ecosystems that can be detected by reduced productivity of desirable plants, undesirable alteration in the biomass and diversity of the micro and macro fauna and flora, accelerated soil erosion, and increased hazards for human occupancy (UNESCO, 1977).

Although there is a substantial base of scientific and technological information on how to control land degradation, the problem is accelerating in many arid and semiarid regions of the world. In the African Sahel, extensive land areas that were once productive have been added to the territories of the Sahara. In the past two decades, deserts have expanded southward in the Sudan by 90–100 km. Large areas of agricultural lands are being degraded in Brazil, Iran, Pakistan, Afghanistan, and the Middle East. Morocco, Algeria, Tunisia, and Libya are also affected (Mageed, 1986).

The FAO (1974) Report on Improving Productivity in Low Rainfall Areas to the Committee on Agriculture identified 64 countries with low-rainfall problems (Table 3). Their report focused on the 21 countries with most of their land (92% average) in low-rainfall areas and with little or no irrigation (category II). A recent study by the U.S. Department of Agriculture (1985) identified 43 developing countries with declining grain production per person, 1950–52 to 1982–84. Of the 21 category II countries mentioned above, 14 showed declines in per capita food production ranging from 2% to 68%, with an average decline of 28% (Table 4).

A recent study (FAO, 1982) of the developing world (excluding east Asia) concluded that with all regions taken as a whole, the total land and water resources were capable of producing sufficient food to sustain twice their 1975 population and one and a half times their projected year 2000 population, even with a low level of agricultural inputs. However, when individual countries were assessed, 15 of 16 countries in southwest Asia were not capable of feeding their present or expected populations from their own lands with low level of inputs, and in Africa, population levels in 30 of 51 countries would exceed the potential supporting capacity of their land resources with low level of inputs by the year 2000. The study further projected that 19 countries will be unable to meet their food needs from national land resources, even with high levels of inputs. These countries, without exception, are dominated by drylands. This study clearly shows the importance of the drylands in many countries and the urgent need for improving the productivity of these areas.

Water resources are very limited in dryland areas, but the principal causes of desertification result from overuse and inadequate management of physical and biological resources. Sung-Chiao (1981) identified resource degradation as a major limitation to productivity in many arid and semiarid regions of China. Conservation of physical resources was identified as the

Table 3. Countries affected by low-rainfall problems[a]

Country	Low-rainfall area (%)	Country	Low-rainfall area (%)	Country	Low-rainfall area (%)	Country	Low-rainfall area (%)
Developing Countries							
Category I[b]:		Category II:		Cateogry III:		Category IV:	
Egypt	100	Botswana	91	Nigeria	28	Angola	23
Iran	85	Chad	92	Argentina	54	Cameroon	9
Iraq	97	Ethiopia	74	Chile	47	Central Afr. Rep.	4
Saudi Arabia	100	Djibouti	100	Mexico	52	Dahomey	24
Yemen (Arab Rep.)	92	Kenya	75	Turkey	41	Ghana	11
Yemen (Dem. Rep.)	100	Mali	95	India	42	Lesotho	20
Pakistan	90	Mauritania	100			Madagascar	15
		Namibia	90			Tanzania	20
		Niger	100			Togo	20
		Senegal	87			Uganda	2
		Somalia	100			Zambia	5
		Spanish Sahara	100			Bolivia	22
		Burkina Faso	94			Brazil	5
		Algeria	96			Columbia	2
		Afghanistan	81			Ecuador	6
		Jordan	98			Paraguay	8
		Libya	100			Peru	17
		Morocco	85			Venezuela	9
		Sudan	91			Lebanon	20[c]
		Syria	83			Sri Lanka	25
		Tunisia	92				

Table 3. Continued

Country	Low-rainfall area (%)	Country	Low-rainfall area (%)	Country	Low-rainfall area (%)
Developed Countries					
Market economies:		Centrally planned economies:			
Australia	82	China	33		
Canada	4	Mongolia	62		
Greece	15	USSR	22		
Israel	75				
South Africa	55				
Spain	33				
USA	35				

Source: FAO (1974).

[a] Some very small countries (Malta, Kuwait, Qatar, Oman, Bahrain) were not included in this list.

[b] Category I: Countries almost entirely dry. Largely dependent on irrigation. Category II: Countries with most of their land in dry areas. Little irrigation. Category III: Countries with appreciable proportion of their land dry. Moderate amount of irrigation. Category IV: Countries with little of their land dry. Little irrigation.

[c] Estimated only. Areas too small to measure with planimeter.

Table 4. Developing countries with declining grain production per person, 1950–52 to 1982–84 (in kilograms per year)

Country	1950–52	1982–84	Decrease (%)
North Africa			
Algeria	219	79	64
Libya	106	69	35
Morocco	258	177	31
Tunisia	196	154	21
Sub-Saharan Africa			
Mozambique	97	36	63
Mali	242	134	45
Angola	81	45	44
Kenya	226	139	38
Nigeria	171	111	35
Ghana	66	44	33
Uganda	155	107	31
Guinea	131	95	27
Rwanda	58	43	26
Zaire	39	32	18
Benin	124	103	17
Senegal	139	118	15
Cameroon	112	97	13
Togo	121	108	11
Liberia	153	139	9
Niger	186	260	9
Sudan	114	104	9
Sierra Leona	155	143	8
Ethiopia	202	189	6
Burkina Faso	181	177	2
Middle East			
Lebanon	54	8	85
Jordan	138	44	68
Iraq	269	105	61
Syria	315	215	32
Iran	193	176	9
Turkey	472	446	5
Latin America			
Haiti	135	75	44
Honduras	194	133	31
Nicaragua	188	136	28
Panama	174	136	22
Chile	192	153	20
Peru	105	85	19
El Salvador	142	129	9
Cuba	55	52	5
Costa Rica	142	141	1

Table 4. Continued

Country	1950–52	1982–84	Decrease (%)
Asia			
Kampuchea	401	267	33
Afghanistan	417	324	22
Nepal	296	243	18
Bangladesh	240	235	2

Source: USDA (1985).

single most important factor for developing sustainable economies in Africa (FAO, 1986).

In a paper prepared for the World Bank, Newcombe (1984) described the stages of land degradation that occur when natural forests are cleared and plowed as people seek new agricultural land. Nutrient cycling is drastically altered, and soil fertility declines following clearing. In the first stage, wood supplies remain plentiful, and gradual erosion is largely unnoticed. As population increases, demand for wood increases for both construction and fuel, initiating a self-feeding cycle of degradation involving cutting wood from remnant forests to generate income, burning crop residues and dung for household fuel, reduced soil fertility, degraded soil structure due to residue and dung removal, and increased vulnerability to wind and water erosion. Eventually, dung and crop residues turn up in markets where only wood was previously sold. As a result of declining organic matter, the cropland becomes less productive, and crop yields prove barely sufficient even for subsistence. Eventually, dung becomes the main fuel source in villages, and rural families use crop residues for cooking and for feeding livestock, which can no longer be supported by grazing land. In the final stage of degradation, crop failures become common even in normal seasons, because topsoil and organic matter depletion have lowered the soil water-holding capacity. As a result, both food and fuel prices rise rapidly. Newcombe (1984) believes a critical point occurs in subsistence economies when more trees are cut for fuel than to make way for farmland.

There is evidence that many regions or countries have entered this cycle of degradation in Africa, Asia, Central America, and South America (Postel, 1984; Brown and Wolf, 1986). A 1980 World Bank study of western Africa showed that fuel wood demand exceeded estimated sustainable yields in 11 of the 13 countries surveyed (Schramm and Jhirad, 1984). India is also facing a fuel wood shortage. Delhi now imports fuel wood from sources 1000 km away (Centre for Science and Environment, 1985), and prices increased at a rate of 4.9% per year from 1977 to 1984 in constant currency (Brown and Jacobson, 1987). A survey in Madhya Pradesh, India, indicated that dung and crop residue exceeded wood as a household fuel by the late 1970s (Centre for Science and Environment, 1985). In western Africa and Central America, as much as 25% of a family's income may go for fuel wood and charcoal (Postel, 1984). It is estimated that at least 400 million tons of manure are burned annually for fuel worldwide (FAO, 1987).

E. Development Objectives

A decrease in land degradation and an increase in the productivity of developing low-rainfall areas can be achieved by improved management of land and water resources. In many locations, improvements can be achieved by more widespread application of known principles of soil and water management to crop and livestock production. In other situations, new concepts and methodologies appropriate to unique aspects of developing areas are required. Lal (1987) presented an excellent review of available low-input technologies that can improve the productivity of dryland regions and protect soil resources from erosion processes. Chase and Boudouresque (1987) described simple use of woodcutters' residues to protect the soil surface, coupled with protection from grazing during the seedling phase, to prevent or arrest degradation of lands following harvest of fuel wood trees from a savanna region in Niger.

Government policies, land tenure arrangements, and social, cultural, and economic factors influence the way in which dryland resources are utilized. When exploring ways to improve natural resource use, the important role played by socioeconomic institutions, in combination with technological advancements, must be recognized. Achieving long-term sustained growth in the productive capacity of low-rainfall areas requires sound decisions based on accurate assessments of resource problems and potentials and on careful analysis of alternative policies, programs, and projects. Hence, the capacity of governments and donor agencies to make technical evaluations and conduct rigorous analyses must be improved if the necessary changes are to occur. A recent study by FAO (1986) outlined specific practices and policies needed to improve African agricultural productivity which focused on provision of incentives, inputs, institutions, and infrastructure.

A critical factor in improving productivity of dryland areas involves implementing improved technology in the field. Policy actions that stimulate the widespread adoption of better soil and water conservation practices, create effective marketing distribution systems for farm commodities, and lead to improved education and information dissemination networks are badly needed for dryland regions. Policies and programs to reduce the rate of population growth must also be developed. Unless current rates of population growth are drastically and immediately curtailed, then all the social and economic goals that governments set for themselves, from higher incomes to food self-sufficiency, will be unattainable or seriously jeopardized in many developing countries.

Overall development goals for improving and sustaining productivity of dryland areas are (1) improve the lives of people who live in dryland areas; (2) improve the contribution that dryland regions make to the growth and development of national economies; (3) sustain the productive life of drylands by arresting the processes of land degradation; (4) rehabilitate land that has already undergone serious degradation; (5) develop systems for dryland management that are economically and sociologically viable and physically sustainable; and (6) improve decision-making ability of national planners.

Technological objectives required to achieve the overall development goals are (1) collect and organize agroclimatic information to form a database sufficient for probabilistic analysis; (2) collect and organize basic soils information required for development of optimal soil and water management practices; (3) collect and organize economic information required to evaluate management practices at the farm, community, region, and national levels; (4) develop sound soil and crop management practices based on proven principles; (5) strengthen multidisciplinary approaches that integrate agroclimatic, soils, agronomic, and economic data for assessment of dryland systems and related policy and program planning; (6) develop demographic and sociological information required for national planning; (7) develop and put in place the infrastructure, such as markets and roads, required for accomplishing the overall goals; and (8) design and put in place social and economic programs to relieve pressure on dryland areas, to foster changes in existing systems, and to restore productivity of degraded resources.

II. Identifying and Alleviating Constraints to Productivity

Dryland farming is a risky enterprise at best. Although a major constraint to dryland agriculture is deficient water, hazards such as insects, diseases, hail, high winds, and intensive rains can destroy crops in a matter of minutes or days. Making matters even more hazardous, farmers in dryland regions are often resource-poor, and these regions are usually of low priority when national resources are allocated. Improving the productivity of dryland regions requires that constraints be clearly identified. Programs then need to be developed that will remove, or at least allow farmers to cope with, these constraints.

Even when there is a knowledge base available for planning and managing crop and livestock systems in dryland regions, the most difficult task is to develop strategies that package technology, necessary infrastructure, and social and economic components together. Perhaps the toughest challenge for both farmers and governments will be to separate measures that are important from those that are expedient. Choosing expedient solutions in dryland areas is always tempting but almost surely leads to failure in the long term. At the same time, the destruction that results from expedient solutions often prevents implementation of the important measures that should have been used initially. An example of an expedient solution is the development of fragile lands for cropland where evidence is clear that cropping cannot be sustained. However, it will become increasingly difficult to avoid misuse of land resources in areas where population pressures are rapidly increasing. Some measures for alleviating known constraints in dryland areas are presented in the following sections.

A. Physical Constraints

1. Agroclimatic Conditions

Crop productivity is a function of the genetic potential of the crop and of the total environment in which the crop is grown. In dryland areas, the environment is often more yield-limiting than the genetic potential of crops. The aerial environment is usually defined in terms of solar radiation, temperature, precipitation, cloudiness, wind, and relative humidity, all of which have a direct effect on the plant. The combination of these factors at a given time and place defines the weather, and the synthesis of weather over long periods of time is climate.

Dominant features of rainfall in dryland regions are its limited amount, temporal and spatial variability, and unpredictibility. With mean annual rainfalls of 200–300 mm, the amount received in a given year ranges from 40% to 200% of the mean, and for areas where the mean is 100 mm per year, the range is from 30% to 350% of the mean (Mageed, 1986). High-rainfall years raise the long-term mean more than low-rainfall years lower it. Consequently, there are more years below the mean than above the mean, with the degree of skewness inversely related to amount of rainfall. This fact makes it imperative that more emphasis be placed on probabilistic analysis, with less attention given to mean values for designing crop calendars. To match a crop to an area, the length of the growing season required for the crop must be matched to periods with favorable temperature for crop growth and a "reasonable" probability of adequate rainfall to produce a yield. Rainfall distribution in space is also very irregular; and, as a result, neighbouring localities can have very different fortunes in the same year. Finally, rainfall intensity is extremely variable, and high-intensity events, even when the amount is relatively low, can result in substantial runoff and soil erosion, so a crop management system must protect the soil resource.

Because of climatic and yield extremes, individuals and governments cannot count on a given production figure for the coming season. Consequently, managing dryland agriculture is, in reality, adopting procedures to cope with the dry, variable climate, including dealing with the economic risks of possible crop failure one year and bumper harvests the next. In either case, farm income may be devastatingly low.

In winter and summer rainfall areas, where there are definite rainy seasons, farmers often respond to a late start in the rainy season by reducing the amount of land sowed or the amounts of seed and fertilizer used. They do this because it is likely that seasonal rainfall will be low if the start of the rainy season is late. On the other hand, they seldom increase inputs significantly in years that the rainy season begins early. Recent studies have shown that seasonal rainfall increases predictably with early onset of the rainy period in areas such as north Africa (Stewart, 1986a), the Near East (Stewart, 1986b), India (Stewart, 1986c), and the Sahelian region (Stewart, 1987). In practical terms, this means that farmers face very different sets of seasonal rainfall probabilities depending on when the rainy season begins. Decisions on crop, plant populations, mineral fertilizers, and other inputs should be based on the beginning of the rainy season. Early onset indicates an increased probability of above-average rainfall, so longer-season crops, higher plant populations, and higher rates of mineral fertilizers should be used. Conversely, the probability of high seasonal rainfall is very low when the rainy season begins very late, and a different set of decisions should be made, often the opposite of those above.

Climatic records of 15–20 years or more, available in many areas, can be used to calculate rainfall probability levels. Rainfall probability analysis should have a very high priority in developing strategies for improving the productivity of drylands. Future emphasis will move toward flexible cropping systems where cropping decisions are based on the availability of stored soil water at certain times and on the probability of growing season rainfall. Computers will become an increasingly important tool in planning and managing successful dryland systems.

Temperature extremes also limit productivity in many dryland areas. Plants are either killed or permanently injured when critical temperatures are exceeded; however, most often the excess heat or cold slows growth rather than being lethal. High temperatures are a growth constraint in the low-altitude tropical and subtropical regions, where the cool season is short or nonexistent. On the other hand, either excessively high or excessively low temperatures can damage crops or limit growth in the higher latitudes, especially in the continental regions. For example, in the northern Great Plains of the United States, loss or damage of fall-sown wheat can occur from extreme cold in the winter, extreme heat (usually combined with dryness) in the summer, or both. In warmer climates, effects of prolonged exposure of crop plants to high soil and air temperatures are not well understood but need to be considered in the development of cropping systems for improved production and water use efficiency.

Soil temperature extremes can often be reduced through simple management practices that alter the properties of the surface layer or the shallow subsoil. For example, mulching the surface with materials such as straw, peat, wood chips, stones, or gravel can markedly reduce soil temperature variations in the shallow layers. A loose soil layer formed by tillage also acts as a thermal blanket and dampens temperature variations in the layers below, compared with unworked soil. These methods can improve moisture conservation and result in more favorable soil temperatures for crop growth and need to be adapted for local situations (Geiger, 1965; Papendick *et al.*, 1973).

Adjusting the crop calendar is another effective method for reducing the risk of damage to crops from either extreme heat or extreme cold. For example, farmers in some areas can delay planting second crops until temperatures moderate after the summer as a means to protect plants against heat injury. In other cases, crop varieties are selected that mature ahead of the hot season, which otherwise would stress plants and reduce crop yields.

Many dryland areas are also subject to winds that adversely affect the water economy of plants, increase soil water evaporation, and cause soil erosion. Wind often accentuates the effect of temperature extremes on plant growth. For example, wind increases heat and cold stress of plants by enhancing desiccation, especially under conditions of low humidity. Gale-force winds that physically damage crops may occur but are, for the most part, isolated weather phenomena and are not a general problem in dryland agriculture. Physical damage to plants caused by blowing sand can be a serious problem in localized areas.

Artificial wind protection has been shown to increase the growth of vegetation and agricultural yields as well as to protect the soil against erosion (Geiger, 1965). Practical methods in dryland areas include planting tree shelterbelts, hedgerows of shrubs, or rows of tall grass or other plants perpendicular to the main wind direction. Alternative windbreaks may also be constructed of materials such as wooden slats, reeds, stones, or combinations of stones and shrubs. The effectiveness of wind barriers depends on the width, height, and density of the belt; crop grown; wind characteristics; and the soil and other local conditions (Geiger, 1965). Other methods to specifically reduce wind erosion include mulching, manipulating field size, soil ridging, interplanting or strip cropping, and tillage to roughen the soil surface after rains (Fryrear and Skidmore, 1985). Keeping the land covered with living or dead vegetation has proved to be among the most effective methods known to control wind erosion (Woodruff *et al.*, 1972).

When forest or grassland areas are being brought into cropland, it is extremely important that attention be given to the relationship of the current weather to the historical range of weather for the area. Catastrophic failures have resulted from converting large areas into cropland during very favorable periods, only to realize later that the years were abnormally wet. Dryland areas are particularly vulnerable, because the soil organic matter declines rapidly after cultivation begins; when droughts occur, the hazard of soil erosion by wind and water becomes severe. A well-known example of such a development is the so-called Dust Bowl of the Great Plains of North America that occurred in the 1930s following an over-optimistic expansion of dryland farming in the 1920s when rainfall was unusually plentiful. The drought of the 1930s led to crop failures and some of the most severe wind erosion ever documented.

2. Soil Characteristics

Soils in the dryland regions of the world range from sandy, shallow, low-fertility soils to highly productive, medium- to fine-textured, deep soils, but the majority of dryland soils have serious problems (Dregne, 1982). Soil characteristics are strongly influenced by the climate in which soils develop, and the interactions of these characteristics with current climatic conditions are a major consideration in understanding the productivity of dryland soils.

Although low soil water levels commonly restrict crop yields in dryland regions, other soil problems such as surface soil hardening, compaction by tillage implements, susceptibility to water and wind erosion, low fertility, shallowness, stoniness, restricted drainage, and salinization also affect crop production. Dregne (1982) developed a generalized map showing principle soil problems in dryland regions of the world (Figure 4).

Increased soil organic matter could alleviate the severity of all the major controllable soil problems, but most practices used in dryland areas tend to reduce organic matter levels. Devising practical and economic means to protect or improve soil resources is a great challenge for researchers, farmers, and policy makers in dryland areas (Dregne, 1982).

The first priority for easing constraints caused by soil problems is to develop a soil capability database and technical and policy plans for sustaining crop production. The most difficult, but also the most vital, part of the strategy will be to withstand the pressure to overdevelop and overuse the resource base, which can make the soil problems more serious through destruction of the resource base.

a. Physical

Poor physical properties of soil are major constraints in dryland regions. Many of the upland soils in the tropical dryland areas are sandy, often gravelly, and generally shallow. These factors contribute to a low water-holding capacity, which makes it more difficult to deal with the detrimental effects of erratic and limited precipitation. Erosion, both wind and water, has intensified these constraints.

There are also large expanses of soils with high clay content. Vertisols are major land resource areas in India, Sudan, Chad, and Australia. While these soils have high water-holding capacities, they have a very limited range of water content and, therefore, of time when they can be tilled. They also require a lot of power for tillage. Management of these soils is particularly difficult when only human or animal power is available.

Soil hardening and crusting are very common in dryland soils and result in large amounts of runoff. When water runs off, there is less water avail-

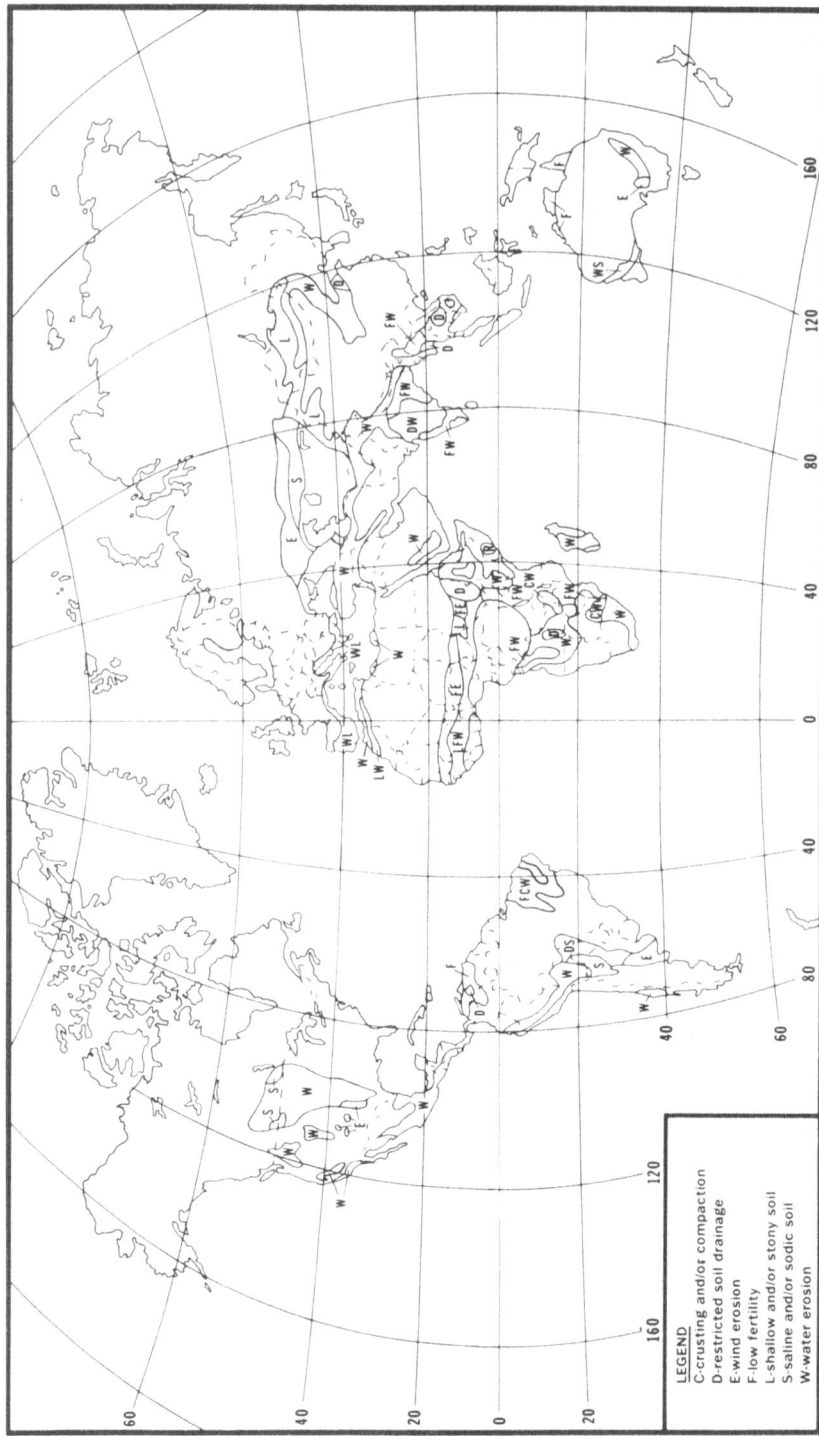

Figure 4. Major soil problems in dryland regions of the world (Dregne, 1982).

LEGEND
C-crusting and/or compaction
D-restricted soil drainage
E-wind erosion
F-flow fertility
L-shallow and/or stony soil
S-saline and/or sodic soil
W-water erosion

able for producing biomass and less input of organic material into the soil, which makes maintenance of good soil physical conditions even more difficult. Tillage is often essential for increasing infiltration (Lal, 1987), but it must be emphasized that tillage should be used as sparingly as possible, because tillage increases the rate of organic matter decomposition. Subsurface tillage, which results in most of the surface residues remaining on the surface, increases infiltration, reduces erosion significantly and results in less organic matter decomposition.

Sweep tillage, often called stubble mulching, was the single most important practice developed to control erosion in the United States Great Plains following the Dust Bowl of the 1930s (McCalla and Army, 1961; Allen and Fenster, 1986). Willcocks (1984), however, cautioned that sweep or chisel tillage would not be satisfactory on very dense soils unless it resulted in adequate loosening of the soil to allow for normal root development of the crop.

Many traditional farming practices exist for rainwater harvesting that could be incorporated into dryland management systems. Use of these techniques helps to minimize the erosive effects of water runoff and increases the water available for crop production (Boers and Ben-Asher, 1982).

b. Chemical

Many dryland soils have serious chemical constraints. Problems include low inherent fertility, acidity, toxic levels of aluminum or other elements, and low nutrient-holding capacity (Dregne, 1982). Extreme spatial variability can make it difficult to obtain research results valid for describing the soil system and developing solutions to relieve constraints (Chase et al., 1987).

Essential plant nutrients can be lost through surface runoff, erosion, leaching, and removal of plant materials. Low inherent levels of phosphate appear to be the major constraint in many Sahelian soils, which greatly reduces the efficiency of use of the limited water available and thus exacerbates the climatic limitation. Results of the Malian-Dutch "Primary Production in the Sahel" project indicate that low fertility is at least as great a limitation to production as water in many of the Sahelian soils (Breman and Uithol, 1987). Soil acidity resulting in aluminum toxicity is a common chemical problem in dryland soils.

Soil testing capabilities should be improved in many developing countries for delineating and addressing these constraints. Strategies should be developed that use combinations of selected crops and cultivars, fertilizers, legumes, and cropping practices to deal with these constraints.

c. Biological

Biological activity in soils is generally much lower in dryland than in more humid zones. The reasons are apparent—lower organic matter levels and periods of extreme dryness. There is also evidence that the organic matter present in dryland soils is chemically and biologically less stable, because there is less biological turnover of organic matter (Anderson, 1987). The net result is that the inherent low level of organic matter in dryland soils decreases more rapidly and to a lower percentage level than in soils of humid zones. The effect is a rapid downward spiral, where the decline in organic matter causes a collapse of the soil structure, a drop in water and nutrient retention, an increase in acidity, and a decrease in plant growth that, in turn, reduces the supply of organic residues (FAO/UNEP, 1983). Forest land, as compared to grassland, has an even more rapid decline rate, because most of the organic matter is at the surface, where decomposition is the most rapid when the land is cultivated. The very rapid and high percentage loss of an already low level of organic matter is the most serious soil problem in dryland regions.

A new equilibrium level of soil organic matter will be reached after new lands are brought into crop production, but that level will be greatly affected by the number and intensity of tillage operations. Any practice that stirs the soil exposes more surface area and increases organic-matter decomposition. Subsurface tillage, which leaves more organic matter on the surface and exposes fewer soil surfaces than most other forms of tillage, results in a higher level of organic matter (USDA, 1974). This improves the overall biological condition of the soil and the chemical and physical characteristics of the soil.

Increasing and maintaining soil organic matter becomes almost impossible when populations of both people and animals require that all the crop residues be utilized for use as fuel or feed. Under extreme conditions, the soil organic matter content drops to the point where the soil becomes useless and soil erosion becomes rampant. This is the final stage in land degradation discussed earlier (Newcombe, 1984).

B. Technological Constraints

1. Soil Fertility

Low native fertility is a widespread problem on sandy soils and on the lateritic ferruginous (iron-rich), medium-textured soils in Africa south of the Sahara, southern and southeastern Asia, northeastern and northern South America, and northern Australia. Aluminum toxicity, the result of extreme soil acidity, is also a major problem in certain locales.

In North Africa, the Middle East, and Africa south of the Sahara, nitrogen and phosphorus deficiencies limit crop production. The lack of some micronutrients is apparent in specific areas, and these deficiencies will intensify and spread as cropping systems intensify. The interactions between nutrients and water are very pronounced, resulting in inadequate response to additional water at low fertility levels and poor response to nutrient additions if water is not available for plant growth. Consequently, nutrient additions must be considered as a component of an improved technology package. A major problem in dryland areas is that single technological improvements may not increase yield, and although total package improvement may be economically beneficial over a period of years, it may not be in low-rainfall years.

The use of soil testing should become more widespread. This technology can be extremely useful in developing countries, because data and experience are not sufficient for making sound fertilizer recommendations. Soil tests can be very helpful in delineating problem areas that suffer phosphorus deficiency and aluminum toxicity. ICARDA (International Center for Agricultural Research in Dryland Areas, Aleppo, Syria) and ICRISAT (International Crops Research Institute in Semi-Arid Tropics, Patancheru, India) have conducted and coordinated studies in recent years, and a database is emerging. Providing plant nutrients in the proper balance is important. Otherwise, farmers may alleviate the limitation of one nutrient only to have other elements become limiting to plant growth.

Biological fixation of nitrogen should be used whenever feasible as a means of improving soil fertility and reducing the need for chemical fertilizers. Incorporation of legume crops into cropping systems is a common way of utilizing biological fixation. It is very important that the legumes be inoculated with the proper *Rhizobium*, especially when a legume is introduced into an area where it was not previously grown, and that adequate phosphorus be available. Gibson *et al.* (1986) reported that the activity of the common diazotroph, *Azotobacter*, could be stimulated by incorporation of wheat straw into the soil to increase the carbon supply available to the bacteria. They believed that up to 50 kg N/ha/year might be added to a soil through enhanced biological fixation under favorable conditions.

Even with the maximization of biological fixation, some fertilizer nitrogen will be required in many situations. Phosphorus, sulfur, and micronutrient deficiencies may have to be overcome by manure or mineral fertilizer additions. Aluminum toxicity problems can be corrected by lime additions, but the best approach will likely be through crop selection and plant breeding for aluminum-tolerant cultivars. Some phosphorus-deficient regions have native rock phosphate resources that may have potential for development which could limit the currency resources required to increase phosphatic fertilization (Kounkandji, 1987).

Mycorrhizal infections can greatly enhance the nutrient-absorbing capacity of root systems in many herbaceous, graminaceous, and tree species, particularly for phosphorus (Read *et al.*, 1985) but also for micronutrients (Killham, 1985). Sieverding (1986) reported that mycorrhiza-infected sorghum plants developed greater root length and were less sensitive to drought stress than noninfected plants because of enhanced phosphorus uptake. Sorghum is a major grain crop in many dryland regions of the world. However, Wang *et al.* (1985) reported that mycorrhizal infection was inhibited at pH below 5.5. Some soils on which plants most need the enhanced nutrient uptake, such as the nutrient-poor, acid Sahelien soils, may be least likely to benefit from use of these organisms. A better understanding of the potential for and limitations to mycorrhizal activity for enhanced nutrient uptake has great potential to improve crop productivity through management of the soil biosphere.

Broadcasting of fertilizer, which is the most common method of application in many regions, may lead to inefficient use of expensive fertilizer inputs. The applicability of new technologies such as improved timing or fertilizer placement between paired rows is unknown. The effect of this technology in "hiding" the fertilizer from access by weeds and decreasing the nutrient immobilization could have significant effects on increasing fertilizer application efficiency, decreasing weed competition, controlling costs, and improving the crop response function (Papendick, 1984).

2. Crop Germplasm

Improved crop varieties and hybrids have resulted in large yield increases in many parts of the world. This factor was a major component of the "green revolution" in Asia, as was an assured moisture supply which reduced risk levels sufficiently to make investments in increased fertilizer and management inputs feasible. The interaction of increased inputs with high-yielding varieties resulted in large payoffs in the classic green revolution wheats and rices. Assured moisture supply is generally lacking in dryland regions, and farmers therefore cannot risk increasing the capital inputs in their crop management systems. Under dryland conditions, the emphasis should be on soil management, especially water-conserving practices, because lack of water is the limiting factor in crop production. More emphasis should also be given to legumes, oilseed crops, and the quality of forage and residue.

In dryland areas where livestock is an important part of the production system, the straw portion of the crops is, in many cases, as important to the farmer as the grain (Cooper *et al.*, 1987). In these regions, increases in grain production due to changing the harvest index are not always desirable. The emphasis under these situations must be to increase biomass. At least in the short run, this will largely be done by improved soil water storage and other soil management practices.

Some attention for long-term plant-breeding programs for improved drought resistance is warranted, but this effort should be a relatively small part of the overall effort, particularly for developing near-term strategies. Improving germplasm for disease and insect resistance is another matter, and this activity and the development of cultivars that are tolerant to aluminum toxicity resulting from soil acidity are extremely important in dryland regions (Wright, 1976). Plants growing in tropical and subtropical areas are subject to a wide variety of disease and insect problems. Major advances have been made in developing improved varieties of major grain crops such as sorghum (House, 1987). Breeding of improved lines of many pulse and root crops merits increased emphasis. Booth *et al.* (1987) describe a climatic analysis technique that can expedite the transfer of plant materials from one region to another, which is especially important when crops are introduced into new areas.

3. Production Practices

Low crop and animal production in dryland farming is not necessarily the result of a lack of scientific knowledge. The principles of dryland farming are fairly well established, and proven practices have been developed for some areas. Although the principles apply worldwide, technologies must be adapted to the local environment as well as to the prevailing social, economic, and institutional conditions. Lal (1987) describes soil management practices that can be incorporated to improve productivity at all levels of agricultural development.

The first priority should be to identify existing technologies and adapt them to specific environments and economic and social conditions. Indigenous technologies should not be overlooked. Where conditions are unique and existing technologies cannot be identified for adaptation, the emphasis should be on developing applied agronomic practices that focus on water conservation and water-fertility interactions, essential steps in increasing biomass production.

a. Agronomic

The most immediate and significant improvements in dryland crop production will result from improved agronomic practices focusing on soil and water conservation, increased biomass production, and cropping systems to maintain soil cover and organic matter levels. In many countries, forage-livestock systems should be given as high a priority as grain production systems.

Agronomic practices are often more effective when two or more are used in concert because of the positive interactions. For example, added fertilizers will not be wholly effective without improved water conservation, and even then they will usually not be beneficial without weed control or if proper nutrient balances are not maintained (Ohm *et al.*, 1985).

Despite its potential, "package" adoption is not evident in the typical sequential adoption pattern of subsistence farmers (Eicher and Baker, 1982). A farmer with limited resources will find it difficult and risky to simultaneously adopt several new techniques that require a shifting of household resources. Moreover, learning a new practice thoroughly may extend over several seasons and have uncertain future payoffs. For these reasons, technology adoption often proceeds slowly despite the potential benefits demonstrated at a research level.

Examples of technologies that could improve the productivity of dryland agriculture include contour ridging, tied ridges, water harvesting, organic and chemical fertilizers, green manures, mulching, weed control, disease and insect control, erosion control practices, and agroforestry/alley cropping.

b. Mechanization and Power

The lack of adequate animal and mechanical traction constrains crop production in many dryland regions. Soil water conditions often change rapidly, and the timing of tillage or other practices can be extremely critical. Improved equipment, ranging from better hand tools to animal-pulled machines to tractor-powered equipment, will be required to make meaningful improvements in production. The size and complexity of equipment must be economically and socially acceptable to the farmers.

Improved tillage systems are urgently needed, because tillage is often desirable to improve infiltration and necessary for weed control (Unger, 1984). However, as already discussed, tillage can be very destructive, because organic matter levels decrease with increased tillage, and this is particularly true in areas of high temperatures (Tate, 1987). As organic matter declines, the soil structure deteriorates, soils become very hard upon drying, and more tillage is necessary to increase water infiltration (FAO/ UNEP, 1983). The organic matter/soil structure/infiltration relationship must be studied, documented, and demonstrated to farmers in these critical soil areas. Tillage, which turns and mixes the surface soil layers, results in rapid degradation of surface residues and soil organic matter. Emphasis should be given to use of implements that undercut the surface without turning the soil. Weed control and improved infiltration will be accomplished by such implements, but erosion will be significantly reduced, and organic matter decomposition will not be as rapid as with surface mixing.

Without question, the most important development in North American dryland agriculture has been maintaining crop residue on the soil surface. The residues control erosion and enhance soil water storage, both of which tend to increase yields. Residue management practices started in Canada in the 1930s with the introduction of V-shaped undercutting implements and has spread throughout the North American drylands. In recent years, chemicals have been widely used for weed control, and tillage is being reduced even more.

Broadcasting is still the dominant practice in many areas for seeding and fertilizing crops. Generally, much more seed and fertilizer are required than for more precise placement. An observed advantage of broadcast seeding, however, is that seeds are often broadcast on ridged land and then the ridges are dragged, resulting in seeds placed from very shallow to fairly deep. If a heavy seeding rate is used, a crop stand is generally assured regardless of the rainfall distribution. A uniform seeding depth sometimes results in failure, because sufficient rain occurs for seed germination, but the crop may be lost before additional rains occur.

Skip cropping, alternating fallow strips with cropped strips, has been used extensively as a means of controlling wind erosion. In some cases, a single row of a tall crop is planted every few meters in a field for wind erosion control. The tall strip crops have the additional advantage of trapping snow and enhancing water storage in colder dryland regions.

c. Fallowing

Fallowing has been used in many agricultural systems to increase the supply of nutrients or water in a soil for production in a subsequent crop. In nutrient-poor soils, fallow has been used primarily to maintain the productivity of the soil over time. In many regions of the world, traditional agriculture has relied on shifting cultivation which allowed long fallow periods for restoration of depleted soils. Areola et al. (1982) reported that soil organic matter, water-holding capacity, and fertility were strongly correlated with the number of years a field had been in fallow, particularly under forest vegetation. However, as lands have come under greater human and animal pressure, fallow periods have been eliminated or shortened, causing severe land degradation (FAO/UNEP, 1983).

In North America, summer fallowing is widely used as a means to increase the water available for succeeding crop growth (USDA, 1974). This practice has been very successful in increasing and stabilizing yields. However, summer fallow has increased wind and water erosion and the decline of organic matter levels.

The basis for summer fallowing can be seen clearly in Figures 5 and 6, showing the effect of increased soil-water storage on grain yield. Substantial amounts of water are required for growing season evapotranspiration before initial grain development, and beyond that, grain yields increase rapidly. Research data and experience in North America suggest that four practices improve water conservation and erosion control in dryland systems: (1) control weeds; (2) leave residues on the surface to reduce evaporation and control erosion and to trap snowfall in some locales; (3) retain hard soil clods 1–8 cm in diameter on soil surface to resist wind erosion, slow runoff water, anchor mulches, provide shade, and physically protect small plants; and (4) manage soil to retain enough water in the seedbed to germinate seeds (Greb, 1979). These conditions are required for both fallow and nonfallow systems.

Conservation tillage systems that leave crop residues on the soil surface are being studied and applied as a means to minimize the adverse effects of summer fallow and to increase the efficiency of soil-water storage. In the early 1900s, when 7–10 tillage operations were common during the fallow period, the percentage of precipitation retained in the soil profile ranged from 18% to 23% in the central Great Plains. Today, with limited tillage

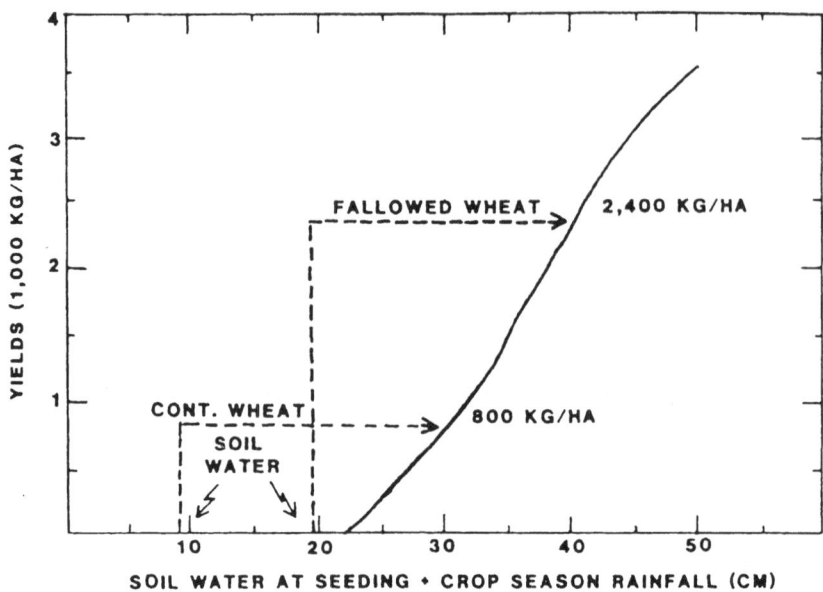

Figure 5. Wheat yield expectancy at North Platte, Nebraska, 1921–1967 (from Greb, 1983, as adapted from Greb *et al.* 1974, and Smika, 1970; reprinted with permission).

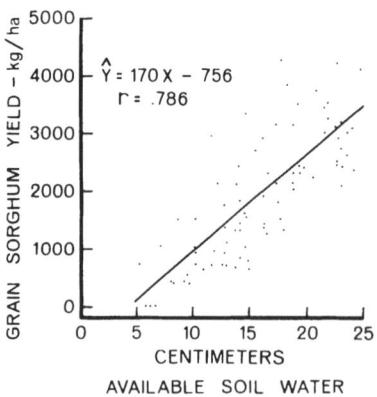

Figure 6. Effect of soil water at seeding (0 to 1.8 m depth) on dryland grain sorghum yield (Jones and Hauser, 1975).

systems, 45–55% efficiencies are achieved. This and other technologies have led to more than a twofold increase in yields.

Greb *et al.* (1979) suggested the following percentage credits: 45% from improved stored water during fallow; 30% to improved wheat varieties with increased harvest index, better tillering, more winterhardiness, earlier ripening to escape heat and hail damage, and more resistance to disease and insect pests; 8% to improved planting equipment; 12% to improved harvesting equipment; and 5% to improved fertilization practices (many soils in the Great Plains have high inherent fertility, particularly under fallow). Of course, the rise in production is due not to any one factor but to the interaction of factors. Elliott and Lynch (1984) reported that straw residues enhance soil aggregation. Lynch (1985) reported that the patterns of N immobilization during straw decomposition could have beneficial effects in that the nitrogen would be immobilized and protected from leaching during the fall and winter months and released in an available form during the spring growth of the crop.

The extent to which fallow is used, or should be used, depends on its effect on quantity, economy, and stability of production and on other factors. These are in turn affected by other considerations, such as soil-water storage capacity, the kind of crops grown, the competition afforded by replacement crops, the type of farming practiced, the weed control needed and afforded, erosion hazards, and the ultimate effect on soil productivity.

The Bushland, Texas, example shown in Figures 1 and 6 illustrates the importance of stored soil water at time of crop seeding in the southern Great Plains of the United States. Since rainfall is significantly less than potential evapotranspiration throughout the year, successful crop production depends on the use of stored soil water in addition to rainfall. Fallowing in other regions, particularly in the winter rainfall regions of the Middle East and north Africa, has not been very successful in increasing stored soil water and subsequent yields (Cooper et al., 1987). Using Amman, Jordan, as an example (Figure 1), a long, dry, hot period follows the rainy season, and very little soil water storage can be maintained through this period. Also, soil water storage prior to seeding is not so important, because the soil profile will be fully recharged in most seasons during the humid portion of the year that occurs soon after seeding crops in the fall. Similar situations occur in summer rainfall regions illustrated by the Rajkot, India, data in Figure 1. Soil water storage is extremely important in these latter two cases for extending the growing season, but they differ greatly from the Bushland, Texas, example in that the soil profiles in those cases are generally fully recharged during the growing season.

Summer fallow was also practiced in Australia in the early 1900s, but some regions experienced yield declines due to low soil fertility associated with the rapid loss of organic matter in the less productive soils. To cope with this problem, Australians incorporated annual self-regenerating species of clovers and medics into cereal crop rotations and integrated cereal and animal production into very successful dryland systems (Puckridge and Carter, 1980). For other regions of the Australian wheat belt where bare fallow systems are still used, Ridge (1986) has shown that fallow enhances the stability of production without reducing the overall productivity of the system.

C. Institutional and Infrastructure Constraints

Dryland farming occupies 97% of the area under cultivation in the Sahel. It also occupies the major land area in northern Africa and the Middle East. Increases in total food production in these areas the past several years have primarily come from expanding the cropland area and reducing the length of fallow. However, per-capita food production is falling, land pressures are becoming more severe, and continued reliance on such extensive means for increasing food production is not realistic. The other option is to implement the institutional and infrastructure changes necessary to become more intensive and raise the level of production.

1. Credit

A move toward more intensive farming systems significantly raises the cost of production and, in dryland areas where moisture supplies are not assured, greatly elevates the risk level of making a profit. In years of severe moisture shortages, the more intensive methods will not increase yields. Since most farmers in these areas have very limited resources, institutional programs must be established to assist the farmer. International donors and governments of developing countries have emphasized subsidized credit to help small farmers adopt novel or risky technologies including natural resource conservation practices. These policies usually did not help small farmers but benefited large, better-off farmers who also had access to these resources (Shaffer *et al.*, 1983). Better institutions in rural areas are needed to ensure that all segments of the communities have access to credit at affordable terms. Local input is essential in the allocation of credit.

Affordable credit must be available to the low-resource farmer if widespread adoption of practices that require expenditures "up front" are to occur. Credit can help the farmer deal with risk associated with changing from traditional practices. Credit must be available in many cases for more than 1 year, because intensive farming systems in dryland areas must have the staying power necessary to take advantage of the favorable years, even though they will often be unprofitable during the less favorable years. At the same time, the systems must provide for family subsistence even in the bad years.

2. Marketing and Distribution

Annual production in dryland regions ranges from very low during years of poor weather conditions to abundant output during years of good rainfall. As discussed above, crop and livestock production systems must take advantage of the favorable years as well as maintain some production and sustain families in the poor years. Effective marketing and distribution systems are essential in this effort. The inability to effectively market produce limits farmers' ability to dispose of surplus output and reduces their income-earning potential. This, in turn, restricts their ability to purchase inputs needed to be fully productive. External inputs are often not available to farmers when they need them at prices they can afford to pay. Moreover, the difficulties farmers face obtaining parts and maintaining equipment create further disincentives for adoption of new technology. As farm incomes remain low, community, regional, and national growth are restricted.

Farm income must rise through marketing before low-resource farmers can invest in the technologies that can enhance the long-term productivity of their land. Farmer cooperatives and trade associations can strengthen small farmers' access to input and output markets.

Most developing regions need improved physical infrastructure. Storage facilities (on-farm and regional) can provide an incentive to produce more than could be consumed by the farm family. All-weather roads, vehicles, and rail systems provide the means to maintain the flow of surplus output to market centers and urban outlets, thus increasing economic activity and opportunity.

Transportation systems also improve access to production inputs, to research and technology transfer services, and to emergency food and fuel during critically bad years. Consequently, infrastructural improvements can lead to substantial productivity gains in dryland areas.

3. Research and Technology Transfer

In dryland regions, especially in many developing countries, research institutions are woefully inadequate. Too often, the resources allocated to drylands have been minimal, because primary attention has been focused on irrigated agriculture or on favorable rainfall areas. Although this past allocation of resources can be easily understood and perhaps even justified, successful development of dryland regions occurs only after research institutions have developed technologies adapted to particular conditions in each area. Data are often inadequate for analyzing agroclimatology and soil resources and management practices. Good databases are essential for the development of dryland regions.

In developing countries, the first priority should be to improve indigenous practices by adapting practices from other regions where such practices have proved successful. The practices from other regions cannot be directly transferred, but the principles will apply, and the specific practices can be altered to fit the local environment and social and economic conditions. The research component should be applied research, focusing on agronomic and cultural practices to increase water conservation and maintain the soil resource.

Economic costs and returns associated with innovative practices must be evaluated. Adaptive research, coupled closely with technology-transfer demonstration plots on the farmers' fields, is essential. Farmers will accept information from other farmers quicker and with more confidence than from technicians. The research must be conducted with equipment appropriate for the region. Research for developing regions must focus on resilient species, minimum effective input rates, low risk of failure, adequate minimum production in poor years, and diverse cropping patterns (Bay-Petersen, 1986). Research and technology transfer programs must be long-term commitments.

4. Fertilizers and Pesticides

Increasing water conservation must be the focal point of improved crop production systems in dryland areas, but water conservation is only the first of several necessary steps. The productivity of many dryland soils cannot be increased without raising the fertility level and controlling pests. The lack of phosphorus is particularly serious in much of Africa and the Middle East. Soil fertility can quickly become the limiting factor in crop production, and the infrastructure (foreign exchange, transportation, and distribution systems) is inadequate in many dryland regions to assure the availability of fertilizers and pesticides. In addition, monitoring and reporting of climatic conditions are often not fed into planning agencies so that limited fertilizer and pesticide inputs can be concentrated in the areas having the most favorable rainfall.

As research results become available regarding the need for fertilizers and pest control, institutional structures must be in place to assure that the information reaches farmers. It is important to ensure that the chemicals are available in areas where they are most needed. Unwise or misguided use of chemical inputs can be very costly and can lead to low efficiency and disenchanted farmers.

5. Farm-Level Knowledge Base

A critical element in technology transfer is the ability of the intended end user to absorb new ideas and effectively utilize new technology over the long run. If dryland farmers are to fulfill their role in the development process, they must become better informed about technical and economic matters that affect them. Traditional land and water management practices generally represent reasonably efficient use of the limited resources available. In the past, these practices were adequate for producing the food and fiber requirements of economies that were either largely self-sufficient or engaged in limited trade with nearby countries. However, traditional methods cannot serve current and future needs to support increased numbers of people.

Despite inherent conservatism and an understandable reluctance to take on risks, typical dryland farmers are receptive to new ideas and interested in technologies that can improve their situation. However, dryland farmers must have a better base of technical knowledge and an understanding of interactions between their farming practices and current and future physical resources. They must also improve their ability to assess new technologies and to keep abreast of economic conditions that affect profitability. In short, the general level of technical and economic knowledge found among dryland farmers must be upgraded if the productivity of the dryland areas of the world is to be raised.

Fundamental needs are better systems of rural education, agricultural production training, and dissemination of information in rural areas. To meet these needs, more and better schools, technology transfer services, producer training programs, field demonstrations, and weather, crop, and marketing reporting services are required.

D. Socioeconomic Constraints

1. Population Growth

During the past few decades, the world has witnessed an unprecedented growth in human population. This has forced equally unprecedented attention to meeting the food, fiber, fuel, and other needs of an expanding population without permitting degradation of soils and other natural resources. Population pressure affects the resource base extensively and intensively. Extensive pressure leads to conversion of grasslands and forests to cropland, with expansion normally progressing into less and less favorable areas. Adequate databases and policies are not available in many cases for making sound decisions regarding expansion of agricultural lands, so land is being destroyed when people indiscriminately cut and burn forests, burn and overgraze grasslands, and cultivate soils that should have remained in native vegetation. The dynamic processes that are being affected on such a massive scale are poorly understood (FAO/UNEP, 1983). This is particularly true in some of the dryland regions of the world where population rates are rapidly rising. Although average population growth in all developing countries peaked at 2.4% a year in 1965 and has since fallen to about 2.1%, the rate of increase is 3% or more and still rising in much of Africa (FAO, 1986). Since many countries in this region depend almost entirely on dryland crop production, enormous pressure is being placed on fragile land resources. With increasing population, millions will settle in drought-prone areas shunned by farmers throughout history. Generally speaking, the drought, in many cases, has not moved to the people; rather, the people have moved and continue to move to drought-prone areas.

Intensive pressure from increased population in dryland regions requires an increase in cropping intensity. The problem now facing many developing countries is that traditional methods of coping with the risks and hazards of dryland agriculture, such as long fallow and nomadism, are breaking down under population pressure, and modern technology has not yet produced acceptable alternatives. The rapid deterioration of soil structure and sharp decline in soil organic matter along with increased weed problems are serious problems in intensive systems that were previously managed by long fallow periods in which bushes or grasses were allowed to grow and decay to restore the productive capacity of the soil.

Dryland areas, as elaborated throughout this chapter, are fragile and very limited in their productive capability even under the best of management. When mismanaged and overused, they become even less productive, resulting in a rapid downward spiral, because less production means fewer root and residue materials for maintaining organic matter. As organic matter declines, the soil becomes very erodible, and severe degradation can occur which renders the land completely useless for plant production. Therefore, the relationship between population levels and productivity in dryland areas is a critical issue that is perhaps the largest of many challenges. Social and economic programs must be developed that address this problem over the long term; otherwise, the resource base in many countries will be damaged beyond repair.

2. Land Tenure and Fragmentation

Land ownership patterns in many parts of the world are based on cultural inheritance traditions and often provide for equal division of agricultural land among heirs. This often results in dividing land into long strips or in small blocks. With small land parcels, use of modern machinery is much more difficult. Weed, insect, and disease control with chemical pesticides is often difficult on small, noncontiguous blocks of land separated by inconvenient distances. Often the land division occurs up and down slope, making it difficult to use sound soil and water conservation practices such as terracing, contouring, and other methods of cross-slope farming. Thus, land fragmentation is a major constraint to improving the productivity of many dryland areas.

Economically sized farms are essential for improvement of land productivity and profitability to farmers. The actual size of an economically viable unit will vary according to availability of water resources, soil characteristics, slope of the land, and climatic features. In developed countries, land holdings have progressively increased in size, whereas in developing countries, farm sizes have become smaller and smaller as they are subdivided among a growing population (FAO, 1987). Potential solutions for land aggregation are to develop national strategies that would encourage combining small parcels into large blocks and greater use of advanced technologies for increasing crop production and conserving soil and water resources. A possibility for land consolidation where family ties are strong is to encourage a system where a single member manages combined family land ownership. Also, in some areas with small landholdings, it may be possible to organize cooperatives and persuade farmers to sow common crops in large blocks to make cultural practices such as water management and pest control possible through cooperative efforts. This has already proved workable in India for some cash crops such as cotton and sugarcane (Papendick *et al.*, 1988). In most situations, aggregating land holdings will be a complex process, because it requires combining technological and social factors to develop solutions.

The first step in dealing with the problem of land fragmentation is to demonstrate clearly that aggregating land parcels increases productivity. Secondly, the constraints to land aggregation must be identified, and potential methods for their removal must be evaluated.

3. Role of Women

Women's role in agriculture has not received adequate attention in agricultural development. Women do more than half the work involved in food production in much of India and Nepal and up to 80% in Africa (Sicoli, 1980). Equipment and related crop production improvements have not been developed to increase the efficiency of women's work because in much of the developing world, women are not typically included in formal and informal training programs or in information networks that lead to better ways of doing things. As a result, technology transfer services, production loans, fertilizer subsidies, and most other support activities are frequently denied to an important human segment of rural areas. Since dryland regions tend to be less advanced than areas where large-scale irrigation programs are in place, it is likely that women in these locations are even more removed from the mainstream of development. It has already been pointed out that lack of land ownership and lack of financing are major constraints to adoption of improved management practices. Women seldom hold title to land or other items of sufficient worth to use as collateral.

Women also have many other labor-intensive tasks necessary for the existence of their families that reduce the time they can spend in agricultural production jobs. As technology packages are developed and strategies are designed for implementing these packages, the role of women will be affected; the success of such programs may very well depend on how successfully their roles are analyzed and addressed.

4. Pastoral Grazing

Livestock play an important role in dryland agriculture and can contribute to more stable income for farm families (Sanford, 1987). However, the numbers of animals frequently exceed the carrying capacity of the land resource. Livestock numbers have grown steadily in recent years, though generally not quite as rapidly as the human population. Dryland areas of developing countries support about 59% of the world's cattle, 82% of buffalo, 41% of sheep, 78% of goats, and 93% of camels (FAO, 1987). A high percentage of these animals are raised in conjunction with farming and are fed on hay, sown forages, and crop residues. Range grazing supports a dwindling share of livestock in industrial and developing countries alike.

Millions of hectares of the world's most fertile grazing lands have been plowed and planted to crops in tandem with accelerating growth of the world's population.

In many parts of the world, nomadic or transhumant grazing is widely practiced in the same manner as it has been for centuries. After crop harvest, fields are grazed by livestock owned or managed by pastoralists. In many cases, the farmers do not have the authority or means to prevent this grazing. Therefore, they do not control this vital portion of the farming system. The removal of both grain and residue from the land results in a rapid decrease in organic matter, soil fertility, and soil structure. With loss in productivity, there is a rapid increase in the likelihood of soil erosion. Before farmers in such areas can institute cropping systems that maintain soil organic-matter levels, cultural and legal factors related to livestock ownership and land utilization must be changed. Policies and programs to deal with this situation will be required before efficient dryland systems can be developed in many regions.

5. Lack of Labor

It is ironic, but true, that in spite of the enormous population growth in many dryland areas, there are often shortages of labor for agricultural production. Shortage of labor at the time of sowing or weeding constrains both the amount of land that can be cultivated by a family and the types of new technologies that can be used effectively. New technologies, if they are to make significant impacts, often must be used as packages of two or more practices such as tied ridges and fertilizer placement. Labor shortages often impede their adoption. A great deal of labor is required from family members (particularly women and children) to meet nonagricultural needs of existence. Two major tasks are often obtaining household water and fuel for cooking. Community development of domestic water supplies and agroforestry projects can reduce the workload of meeting these needs and increase the availability of family labor for farming operations.

The timing of operations in dryland farming systems is critical. Proper timing is essential for water conservation, seeding, weed control, and erosion control. A matter of a few days, and in some cases hours, can mean the difference between success and failure. Seeding operations are particularly critical in a summer or winter rainfall area with a limited growing period, such as the season shown in Figure 1 for Rajkot, India. Delays of even a few days will limit the yield of the crop, because the rains stop very abruptly, and crop growth cannot be extended beyond the length of time it takes to deplete the soil water reserve. In continental climates, such as represented by Bushland, Texas (Figure 1), there is the possibility of precipitation at any time of the year, so a delay in planting is not necessarily associated with a reduced yield in any given year.

Since labor shortages during crop production and harvest periods play important roles in determining what crops are produced and in what quantities, policy measures must be developed that target this important issue. Actions to remove rural/urban wage rate disparities, improve seasonal labor mobility from region to region (including the urban unemployed and imported labor), and improve rural living conditions for migratory workers are universally needed to help solve labor shortages in remote dryland areas.

6. Macroeconomic Policy Constraints

To a great extent, the macroeconomic policies of a country can either augment or diminish programs designed to stimulate agricultural production and reduce rural poverty. In general, the economic policies of developing countries in past years have had negative effects on development in the dryland regions. Development strategies have shifted resources away from dryland to irrigated production and from rural to urban areas. Since dryland farmers are poorer and politically less influential, the effects of adverse macroeconomic policies fall disproportionately on them, in spite of the fact that they are often the primary producers of food crops.

Examples of macroeconomic policies that have hindered development of stable dryland agricultural systems include (1) disparity in urban-rural wage rates, which draws labor from agricultural areas; (2) government reliance on food imports and food aid, leading to depressed domestic food prices and reduced farmer income (OTA, 1986); (3) an overvalued exchange rate, which reduces export potential and encourages imports (Timmer et al., 1983); (4) export taxes on agricultural commodities, which further exacerbates the income-earning problem faced by farmers; (5) cheap food policies that benefit urban consumers, creating severe problems for food producers; (6) unchecked inflation, which together with commodity price ceilings further erodes the incentive to invest in improved technology; and (7) taxation and subsidy policies, which lead to distorted prices for inputs. Subsidies are often designed to offset a tax or to adjust for commodity price manipulations, but these subsidies are generally ineffective in reaching the low-resource farmer. Subsidies usually lead to shortages and possibly rationing, with the result that the relatively well-off farmers benefit most. Good examples of these problems are government-subsidized credit and fertilizer.

When governments place too much emphasis on developing irrigation schemes, generally to increase production of export crops, limited public resources are diverted to the costly dams, canals, infrastructure, and farm production systems necessary to make irrigation possible. This means that fewer resources are available to help solve dryland problems (Lele, 1984). Since irrigated farming tends to be highly subsidized and supported, it also means that the dryland farmer is at an economic disadvantage in competing for markets, inputs, and new technology. Alleviating these constraints means modifying policies that have negative impacts on dryland agriculture or implementing new ones effective in bringing about needed changes.

Policies should bring input-output price ratios more in line with market supply and demand, including trade in international markets. At the same time, greater effort is needed to find ways in which society at large can share with dryland producers the risks of climatic variability and the ups and downs of domestic and international trade through appropriate government policies and programs. This kind of support is particularly important in the early stages of development of dryland agriculture as it becomes more dependent on the marketing of inputs and outputs. Price stability programs, therefore, should be given high priority.

Government and donor agencies should also strengthen effects to encourage more widespread adoption of improved soil and water management practices in dryland areas. For example, tax advantages tied to conservation investments can provide incentives for farmers to carry out costly, long-term land improvements that might not otherwise occur. Underwriting part of the cost of inputs and specialized equipment—e.g., tied ridgers—can do the same. Improving agricultural extension and soil and water conservation technical support services is a policy decision that must be taken.

In summary, the macroeconomic policies developing countries follow have far-reaching impacts on dryland agricultural productivity and on the overall growth and development of dryland regions. The macropolicy climate must be conducive to change and technological advancement. Macropolicies do this by creating an environment in which the appropriate incentives are allowed to operate and where government efforts are directed to all sectors of the agricultural economy, dryland and irrigated alike.

III. Conclusions

As an increasing number of people live and support themselves in dryland regions of the world, the stress placed on fragile dryland systems is becoming more and more evident. Inadequate and erratic crop production in these regions leads to tremendous hardship for the people dependent on dryland agriculture. Governments of many developing countries are being asked to deal simultaneously with food relief, the need for rapid and sustained agricultural production increases, protection of physical and biological resources from degradation, and reclamation of degraded lands. Achieving developmental objectives, which often seem conflicting, requires sound planning and large investments of time and money. However, the urgent need for rapid development of dryland regions, which are essential to the food and fiber production of so many countries in the world, can no longer be overlooked.

Soil and water management practices that can result in increased productivity over a relatively short term with few purchased inputs include contour terracing, tied ridges, water harvesting, and improved weed control. To do this, on-farm demonstrations and adaptive research need to be strengthened, and policies to relieve labor bottlenecks must be implemented. Over the medium term, it will be essential to identify, adapt, develop, and package practices that can further increase agricultural productivity through the use of purchased inputs, such as fertilizers, improved tools and implements, and improved seed varieties with disease and insect resistance. Technological recommendations must allow farmers flexibility in choosing production practices that best fit their particular environment. To identify reasonable technologies to meet the needs of specific regions, detailed databases for agroclimate, soil, and other resources must be developed for planning purposes.

The fragility and importance of the soil and water resource base must receive greater recognition, at both the national and on-farm levels. Productivity cannot be sustained if the resources that support productivity are degraded. Governments must develop criteria to detect the rate of land degradation, to evaluate the impact of degradation on productivity, and to determine the population-supporting capacity of their land. Implementation of policies that will ensure that lands are not expected to produce beyond their capacity is a major long-term challenge.

National policies must recognize that contributions of dryland agriculture are vital to the national economy and to the well-being of the people living in the dryland regions. National tax, currency, and wage/price policies must treat the dryland agricultural regions fairly. Dryland farmers cannot assume all of the risks of production in extremely risky environments if access to stable markets, credit, storage, and purchase of food are not assured.

Improved cropping systems must evolve from the agricultural and social systems currently in place, but the importance of resource protection and maintenance must be recognized. Development and maintenance of improved agricultural productivity will require governments to strengthen rural area infrastructure and improve national economic policies. Development will also require institutions at the local and community level that can interact directly with farmers to bring about better land and resource management. Without effort at all levels, the situation of the people in many dryland areas will continue to deteriorate.

Acknowledgments

The authors acknowledge the important contribution to this paper by Dr. B.A. Stewart. Preparation of this review would not have been possible without his generous sharing of time, information, and ideas.

References

Allen, R.R., and C.R. Fenster. 1986. Stubble-mulch equipment for soil and water conservation in the Great Plains. *J. Soil Water Conserv.* 41:11–16.

Anderson, D.W. 1987. Pedogenesis in the grassland and adjacent forests of the Great Plains. In: B.A. Stewart (ed.), *Advances in Soil Science*, Vol. 7. Springer-Verlag, New York, pp. 53–93.

Areola, O., A.O. Aweto, and A.S. Gbadegesin. 1982. Organic matter and soil fertility restoration in forest and savanna fallows in southwestern Nigeria. *Geo-Journal* 6:183–192.

Bay-Petersen, J. 1986. New technology for low-income farmers in developing countries. *Outlook Agric.* 15:110–114.

Boers, T.M., and J. Ben-Asher. 1982. A review of rainwater harvesting. *Agric. Water Manag.* 5:145–158.

Booth, T.H., H.A. Nix, M.F. Hutchinson, and J.R. Busby. 1987. Grid matching: A new method for homoclime analysis. *Agric. Forest Meteor.* 39:241–255.

Breman, H., and P.W.J. Uithol. 1987. Communication of the results of the "Primary Production in the Sahel (PPS)" research project. *Soil Tillage Res.* 9:387–393.

Brown, L.R., and J. Jacobson. 1987. Assessing the future of urbanization. In: L.R. Brown (proj. dir.), *State of the World 1987, A Worldwatch Institute Report on Progress Toward a Sustainable Society*. W.W. Norton, New York, pp. 38–56; 219–222.

Brown, L.R., and E.C. Wolf. 1986. Assessing ecological decline. In: L.R. Brown (proj. dir.), *State of the World 1986, A Worldwatch Institute Report on Progress Toward a Sustainable Society*. W.W. Norton, New York, pp. 22–39, 215–217.

Centre for Science and Environment. 1985. *The State of India's Environment*. SCE, New Delhi, India.

Chase, R.G., and E. Boudouresque. 1988. A study of methods for the revegetation of barren, crusted Sahelian forest soils. In: C.M. Renard (ed.), *Proc. Int. Workshop on Soil, Water and Crop/Livestock Management Systems for Rainfed Agriculture in the Sudano-Sahelian Zone*, Niamey, Niger, January 1987. ICRISAT Center, Patancheru, India, p. 115.

Chase, R.G., J.W. Wendt, and L.R. Hossner. 1988. Crop growth variability in sandy Sahelian soils. In: C.M. Renard (ed.), *Proc. Int. Workshop on Soil, Water and Crop/Livestock Management Systems for Rainfed Agriculture in the Sudano-Sahelian Zone*, Niamey, Niger, January 1987. ICRISAT Center, Patancheru, India, p. 126.

Cooper, P.M.J., P.J. Gregory, D. Tully, and H.C. Harris. 1987. Improving water-use efficiency of annual crops in the rainfed farming systems of west Asia and north Africa. *Exp. Agric.* 23:113–158.

Doorenbos, J., and W.O. Pruitt. 1977. Crop water requirements. FAO Irrigation and Drainage Paper No. 24. FAO, Rome.

Dregne, H.E. 1982. Dryland soil resources. Science and Technology Agriculture Report, Agency for International Development, Washington.

Eicher, C., and D. Baker. 1982. Research on agricultural development in sub-Saharan Africa: A critical survey. International Development Paper No. 1, Michigan State University, East Lansing.

Elliott, L.F., and J.M. Lynch. 1984. The effect of available carbon and nitrogen in straw on soil aggregation and acetic acid production. *Plant Soil* 78:335–343.

FAO. 1974. Improving productivity in low rainfall areas. Report to the Committee on Agriculture. COAG/74/4 Rev. 1. FAO, Rome.

FAO. 1978a. Report on the agro-ecological zones project: Methodology and results for Africa. *World Soil Resource Report 481*. FAO, Rome.

FAO. 1978b. Report on the agro-ecological zones project: Results for southwest Asia. *World Soil Resource Report 48/2*. FAO, Rome.

FAO. 1980. Report on the agro-ecological zones project: Results for southeast Asia. *World Soil Resource Report 48/4*. FAO, Rome.

FAO. 1981. Report on the agro-ecological zones project: Methodology and results for South and Central America. *World Soil Resource Report 48/3*. FAO, Rome.

FAO. 1982. Potential population supporting capacities of lands in the developing world. *Report FPA/INT/513*. FAO, Rome.

FAO. 1986. African agriculture: The next 25 years. *Main Report ARC/86/3*. FAO, Rome.

FAO. 1987. Improving productivity of dryland areas. *Committee on Agriculture Report COAG/87/7*. FAO, Rome.

FAO/UNEP. 1983. *Guidelines for the Control of Soil Degradation*. FAO, Rome.

Fryrear, D.W., and E.L. Skidmore. 1985. Methods for controlling wind erosion. In: R.F. Follett and B.A. Steward (eds.) *Soil Erosion and Crop Productivity*. ASA-CSSA-SSSA. Madison, WI, pp. 443–457.

Geiger, R. 1965. *The Climate Near the Ground*. Harvard University Press, Cambridge, MA.

Gibson, A.H., M.M. Roper, and D.M. Halsall. 1986. Straw breakdown to fuel nitrogen fixation in soil. *CSIRO Div. Plant Ind. Rep. 1985–86*. CSIRO Inst. Bio. Res., Canberra, Australia.

Greb, B.W. 1979. Technology and wheat yields in the central Great Plains: Commercial advances. *J. Soil Water Conserv.* 34:269–273.

Greb, B.W. 1983. Water conservation: Central Great Plains. In: H.E. Dregne and W.O. Willis (eds.) *Dryland Agriculture*, Agronomy monograph 23. ASA-CSSA-SSSA, Madison, WI, pp. 57–72.

Greb, B.W., D.E. Smika, and J.R. Welsh. 1979. Technology and wheat yields in the central Great Plains: Experiment station advances. *J. Soil Water Conserv.* 34:264–268.

Greb, B.W., D.E. Smika, N.P. Woodruff, and C.J. Whitefield. 1974. Summer fallow in the central Great Plains. In: *USDA Conserv. Res. Rep. No. 17*. USDA, Washington, pp. 51–85.

House, L.R. 1987. Sorghum—present status and future potential. *Outlook Agric.* 16:21–27.

Jones, O.R., and V.L. Hauser. 1975. Runoff utilization for grain production. In: *Proc. Water Harvest Symposium, March 26–28, 1974, Phoenix, AZ*. ARS-W-22, USDA, Washington, pp. 277–283.

Killham, K. 1985. Vesicular-arbuscular mycorrhizal mediation of trace and minor element uptake in perennial grasses: Relation to livestock herbage. In: A.H. Fitter (ed.), *Ecological Interactions in Soil*. Special Publ. No. 4, British Ecol. Soc. Blackwell Scientific Publications, Oxford, U.K., pp. 225–232.

Koundandji, E. 1987. Use of natural phosphates in low rainfall zones. In: C.M. Renard (ed.), *Proc. Int. Workshop on Soil, Water and Crop/Livestock Management Systems for Rainfed Agriculture in the Sudano-Sahelian Zone*, Niamey, Niger, January 1987. ICRISAT Center, Patancheru, India, p. 130.

Lal, R. 1987. Managing the soils of sub-Saharan Africa. *Science* 236:1069–1076.

Lele, U. 1984. The role of risk in an agriculturally led strategy in Sub-Saharan Africa. *Am. J. Agric. Econ.* 66(5):677–683.

Lynch, J.M. 1985. Microbial saprophytic activity on straw and other residues: Consequences for the plant. In: R.H. Fitter (ed.) *Ecological Interactions in Soil.* Special Publ. No. 4, British Ecol. Soc. Blackwell Scientific Publications, Oxford, U.K., pp. 181–191.

Mageed, Y.A. 1986. *Anti-Desertification Technology and Management.* United Nations Environment Programme, Nairobi, Kenya.

McCalla, T.M., and T.J. Army. 1961. Stubble mulch farming. *Adv. Agron.* 13:125–196.

Newcombe, K. 1984. An economic justification for rural afforestation: The case for Ethiopia. *Energy Dept. Paper No. 16.* World Bank, Washington, D.C.

OTA (Office of Technical Assistance), U.S. Congress. 1986. *Continuing The Commitment: Agricultural Development in the Sahel.* U.S. Government Printing Office, Washington.

Ohm, H., J. Nagy, and S. Sawadogo. 1985. Complementary effects of tied ridging and fertilization with cultivation by hand and donkey and ox traction. In: H. Ohm and S. Nagy (eds.), *Appropriate Technologies for Farmers in Semi-Arid West Africa.* Purdue University Press, Lafayette, IN. pp. 61–73.

Oram, P. 1980. What are the world resources and constraints for dryland agriculture? In: *Proc. Int. Congress Dryland Farming.* South Australia Department of Agriculture, Adelaide, Australia, pp. 17–78.

Papendick, R.I. 1984. *Tillage Practices in Low Rainfall Agriculture in the United States.* FAO Panel of Experts on Agricultural Mechanization, Report of the 6th Session. University of Cukurova, Adena, Turkey. FAO, Rome.

Papendick, R.I., M.J. Lindstrom, and V.L. Cochran. 1973. Soil mulch effects on seedbed temperature and water during fallow in eastern Washington. *Soil Sci. Soc. Am. Proc.* 37:207–314.

Papendick, R.I., F.L. Choudhury, and C. Johnson. 1988. Managing systems for increasing productivity pulses in dryland agriculture. In: R.J. Summerfield (ed.) *World Crops: Cool Season Food Legumes.* Martinus Nijhoff/Dr. W. Junk, Dordrecht, The Netherlands (in press).

Postel, S. 1984. Protecting forests. In: L.R. Brown (proj. dir.), *State of the World 1984, A Worldwatch Institute Report on Progress Toward a Sustainable Society.* W.W. Norton, New York, pp. 74–94, 219–222.

Puckridge, D.W., and E.D. Carter. 1980. Dryland agriculture in southern Australia and its relationship to other regions—a perspective. In: *Proc. Int. Congress Dryland Farming.* South Australia Department of Agriculture, Adelaide, Australia, pp. 155–190.

Read, D.J., R. Francis, and R.D. Finaly. 1985. Mycorrhizal mycelia and nutrient cycling in plant communities. In: A.H. Fitter (ed.) *Ecological Interactions in Soil.* Special Publ. No. 4, British Ecol. Soc. Blackwell Scientific Publications, Oxford, U.K., pp. 193–217.

Ridge, P.E. 1986. A review of long fallows for dryland wheat production in southern Australia. *J. Aust. Inst. Agric. Sci.* 52:37–44.

Sanford, S. 1987. Crop residue/livestock relationships. In: C.M. Renard (ed.), *Proc. Int. Workshop on Soil, Water and Crop/Livestock Management Systems for Rainfed Agriculture in the Sudano-Sahelian Zone,* Niamey, Niger, January 1987. ICRISAT Center, Patancheru, India, p. 148.

Schramm, G., and D. Jhirad. 1984. *Sub-Saharan Africa Policy Paper—Energy*. World Bank, Washington.

Shaffer, J., M. Weber, M. Riley, and J. Staatz. 1983. Influencing the design of marketing systems to promote development in Third World countries. Paper presented at Int. Workshop Agric. Markets in the Semi-Arid Tropics, ICRISAT, Hyderabad, India.

Sicoli, F. 1980. Women in rural development: Recommendations and realities. *Ceres* (FAO Review on Agriculture and Development) 13(3):15–22.

Sieverding, E. 1986. Influence of soil water regimes on VA mycorrhiza. IV. Effect on root growth and water relations of sorghum bicolor. *J. Agron. Crop Sci.* 157:36–42.

Smika, D.E. 1970. Summer fallow for dryland winter wheat in the semi-arid Great Plains. *Agron. J.* 62:15–17.

Stewart, B.A., and E. Burnett. 1987. Water conservation technology in rainfed and dryland agriculture. In: W.R. Jordon (ed.) *Water and Water Policy in World Food Supplies*. Texas A & M University Press, College Station, pp. 355–359.

Stewart, J.I. 1986a. Mediterranean-type climate, wheat production and response farming. In: R.I. Papendick, J.F. Parr, and C.E. Whitman (eds.), *Proc. Workshop on Soil, Water and Crop/Livestock Management Systems for Rainfed Agriculture in the Near East Region*, Amman, Jordan, January, 1986. USDA, Washington (in press).

Stewart, J.I. 1986b. Response farming for improvement of rainfed crop production in Jordon. In: R.I. Papendick, J.F. Parr, and C.E. Whitman (eds.), *Proc. Workshop on Soil, Water and Crop/Livestock Management Systems for Rainfed Agriculture in the Near East Region*, Amman, Jordan, January 1986. USDA, Washington (in press).

Stewart, J.I. 1986c. Development of management strategies for minimizing the impact of seasonal variation in rainfall. In: *Proc. Int. Consultants' Meeting on Research on Drought Problems in the Arid and Semi-Arid Tropics*, Hyderabad, India, November 1986. ICRISAT Center, Patancheru, India (in press).

Stewart, J.I. 1987. Potential for response farming in sub-Saharan Africa. In: C.M. Renard (ed.), *Proc. Int. Workshop on Soil, Water and Crop/Livestock Management Systems for Rainfed Agriculture in the Sudano-Sahelian Zone*, Niamey, Niger, January 1987. ICRISAT Center, Patancheru, India, p. 120.

Sung-Chiao, 1981. Desert lands of China. I. The sandy deserts and the Gobi: A preliminary study of their origin and evolution. ICASALS Publ. No. 81–1. Int. Center for Arid and Semi-Arid Land Studies. Texas Tech University, Lubbock.

Tate, R.L. 1987. *Soil Organic Matter: Biological and Ecological Effects*. John Wiley, New York.

Timmer, C., W. Falcon, and S. Pearson. 1983. *Food Policy Analysis*. Johns Hopkins University Press, Baltimore.

UNESCO. 1977. *World Map of Desertification*. United Nations Conference on Desertification Report A/Conf. 74/2. United Nations, New York.

Unger, P.W. 1984. *Tillage Systems for Soil and Water Conservation*. FAO Soils Bulletin 54. FAO, Rome, 278 pp.

USDA, Agricultural Research Service. 1974. Summer fallow in the western United States. Conservation Research Report No. 17. USDA, Washington, 160 pp.

USDA. 1985. *World Indices of Agricultural and Food Production, 1950–84*. Economic Research Service (unpublished printout), Washington.

Wang, G., D.P. Stribley, P.B. Tinker, and C. Walker. 1985. Soil pH and vesicular-arbuscular mycorrhizas. In: A.H. Fitter (ed.) *Ecological Interactions in Soil*. Special Publ. No. 4, British Ecol. Soc. Blackwell Scientific Publications, Oxford, U.K., pp. 219–224.

Willcocks, T.J. 1984. Tillage requirements in relation to soil type in semi-arid rainfed agriculture. *J. Agric. Eng. Res.* 30:1–10.

Woodruff, N.P., L. Lyles, F.H. Siddoway, and D.W. Fryrear. 1972. How to control wind erosion. *USDA Agric. Info. Bull. 354*, USDA, Washington, 22 pp.

Wright, M.J. 1976. *Plant Adaption to Mineral Stress in Problem Soils*. Spec. Pub., Cornell Univ., Agric. Exp. Sta., Ithaca, NY.

A Conceptual Model of Changes in Soil Structure Under Different Cropping Systems

R.J. Gibbs* and J.B. Reid[†]

I. Introduction

There is much evidence that soil structure can be markedly affected by a farmer's choice of cropping systems. For example, tillage type and intensity may substantially influence soil structure (for reviews see Cannell, 1985; Davies *et al.*, 1982; R.S. Russell, 1977). The choice of crop may also result in significant alterations to soil structure. Many researchers have found deteriorations in soil structure under arable crops such as corn (*Zea mays*) (Page and Willard, 1946), potatoes (*Solanum tuberosum*) (Doyle and Hamlyn, 1960), wheat (*Triticum aestivum*) and barley (*Hordeum vulgare*) (Low, 1972). Improvements in soil structure have been observed under a

*Present address: Soil Science Unit, University College of Wales, Penglais, Aberystwyth, Wales SY23 3DE.
[†]Present address: MAFTech, Ministry of Agriculture and Fisheries, P.O. Box 1654, Palmerston North, New Zealand.

© 1988 by Springer-Verlag New York Inc.
Advances in Soil Science, Volume 8

number of ley and pasture grasses (see e.g., A.L. Clarke *et al.*, 1967; Robinson and Jacques, 1958) and alfalfa (*Medicago sativa* L.) (Cooke and Williams, 1972). Figure 1 illustrates some measured changes in soil structure which can be attributed to different cropping systems. Harris *et al.* (1966) and E.W. Russell (1973) summarized much of the earlier research in this area.

However, the apparent general contrast between the effects of arable and grass crops, as indicated by the literature, cannot necessarily be relied on in specific instances. In our experience, there are many farmed sites where soil structure remains poor despite long periods under grass. Conversely, many long-term arable soils may retain good soil structure (A.L. Clarke and Russell, 1977; Ross and Hughes, 1985; Lance, 1987). Clearly, factors other than the choice of crop species must also be considered when recommending means of ameliorating or maintaining soil structure. Similar complexities exist when deciding tillage regimes on a site-by-site basis (Cannell, 1985).

In many instances, changes in soil structure caused by cropping systems are very important because these changes strongly influence crop growth, the potential for soil erosion, and the ease of land and crop management (especially under wet conditions). Mathematical models are now becoming available to describe the consequences of structural differences for soil processes, crop growth, and land management (Unger and Van Doren, 1982). However, to use these models to their fullest potential requires the ability to predict how and when soil structure might change.

A. Problems to be Addressed

Despite the large number of publications on the subject, it remains difficult to predict, or even interpret, changes in soil structure in any particular locality. The many possible combinations of soil type, crop, management and weather often seem confusing enough, but two further problems compound these difficulties.

First, comparisons between reported experiments are complicated by the great variety of techniques used by researchers to assess soil structure. Many of these techniques are quite arbitrary and yield little information that can be used for quantitative predictions directly relevant to crop growth or management. An illustration of this is the abundance of different methods used to determine aggregate stability.

Second, several of the processes known to affect soil structure may themselves be influenced by a change in cropping system, and many of these processes interact strongly. Examples include root activities (Tisdall and Oades, 1979; Reid and Goss, 1981), earthworm activities (Douglas *et al.*, 1980/81; Syers and Springett, 1984) and tillage (Davies *et al.*, 1982). Relatively few field experiments have attempted to identify the contributions

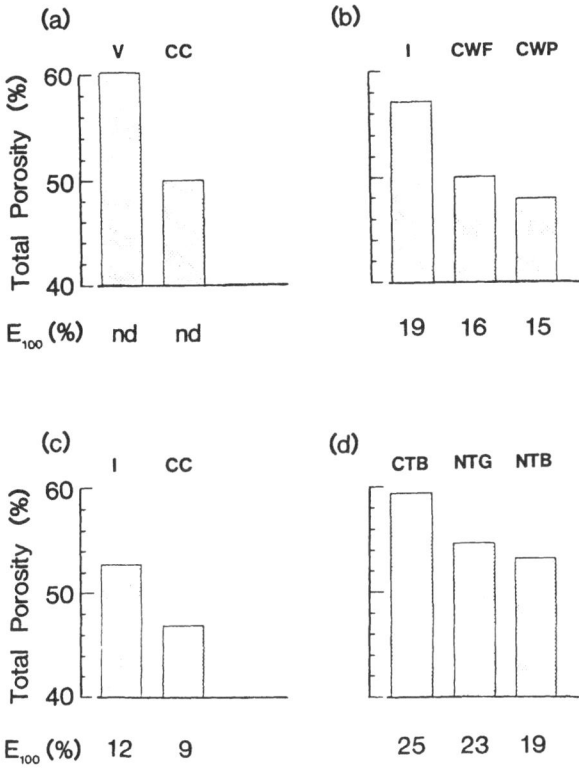

Figure 1. Effects of cropping systems upon soil total pore space and macropores >100 μm equivalent diameter (E_{100}). (a) Nappanee silty clay loam, Ohio. Measurements made at 0–30 cm depth. V = virgin, uncultivated woodland; CC = land cultivated for corn production over 40 years (tillage included plowing); nd = not determined. (From Page and Willard, 1946.) (b) Walla-Walla silt loam (Typic Haploxerolls), Oregon. Measurements were made at 10 cm depth after wheat harvest. I = idle land (uncultivated for >59 years, vegetation miscellaneous weeds and *Bromus tectorum*); CWF = cultivated land under wheat–summer fallow rotation for >50 years (tillage included moldboard plowing); CWP = cultivated land under wheat and peas (*Pisum sativum*) rotation for >50 years (tillage included moldboard plowing). (Calculated from Allmaras *et al.*, 1982.) (c) Ironside sandy loam (Gleyed Melanic Brunisol), Ontario, Canada. Measurements were made at 0–10 cm depth in August and September. I = idle land (uncultivated for >35 years, vegetation grass and some trees; CC = land cultivated for corn production >35 years. (Calculated from Coote and Ramsey, 1983.) (d) Mangateretere silt loam (Typic Haplaquept), Hawkes Bay, New Zealand. Samples were taken at 5–15 cm depth, 4 days after spring cultivation. CTB = tilled land in summer tomatoes (*Lycopersicon esculentum*)–winter beans (*Vicia faba*) rotation (tillage included rotary hoeing); NTG = untilled land in summer tomatoes–winter grass (*Lolium multiflorum*) rotation; NTB = untilled land in summer tomatoes–winter beans rotation. All treatments were in their fourth year. (Unpublished results, J.B. Reid.)

made by the various processes responsible for structural changes under different cropping systems. Nevertheless, such experimentation is necessary if we are to order our knowledge of cropping system effects so that reliable predictions can be made. There is therefore a need for a realistic conceptual model that (1) summarizes current knowledge of the mechanisms reponsible for cropping system effects on soil structure, (2) provides a satisfactory framework for the design and interpretation of experiments concerning the effects of cropping systems on soil structure, and (3) has the potential to be developed mathematically for predictive purposes. The objective of this paper is to describe and discuss such a model.

II. The Conceptual Model

A. General Approach

The model concerns mainly the creation and destruction of soil macropores, because macropores strongly influence movement of roots, air, water, and solutes through soil (Barley, 1954; E.W. Russell, 1971; Cary and Hayden, 1973; R.S. Russell and Goss, 1974; Greenland, 1977; Goss and Reid, 1981; Germann and Bevan, 1981; Scotter and Kanchanasut, 1981; Klute, 1982; Goss et al., 1984). Furthermore, many of the most significant effects that cropping systems have on soil physical processes appear to be due to changes in soil macroporosity (A.L. Clarke et al., 1967; Douglas et al., 1980/81; Allmaras et al., 1982; Bridge et al., 1983; Coote and Ramsey, 1983; Ahuja et al., 1984; Ross and Hughes, 1985). Tables 1 and 2 give some indication of how soil management may affect macroporosity and hence soil transmission properties. It is worth noting from Table 2 how tillage increased the total and macroporosity but resulted in less continuous

Table 1. Saturated hydraulic conductivity (K_{sat}) and soil porosity at 0–10 cm depth in Ironside sandy loam (Gleyed Melanic Brunisol)

	$10^5 \times K_{sat}$ (m s^{-1})	E_t (%)	E_{100} (%)	E_{50} (%)	E_{30} (%)
Idle land[a]	6.2	53	12	17	19
Cultivated for corn[a]	1.3	47	9	13	16

Source: Calculated from results of Coote and Ramsey (1983).

E_t = % of soil volume occupied by pores (i.e., total porosity).
E_{100} = % of soil volume occupied by pores >100 μm equivalent diameter. E_{50} = % of soil volume occupied by pores >50 μm equivalent diameter. E_{30} = % of soil volume occupied by pores >30 μm equivalent diameter.
[a]See Figure 1c.

Table 2. Saturated hydraulic conductivity (K_{sat}) and soil porosity at 20–30 cm depth in Denchworth series clay soil (Aeric Haplaquept).

	$10^5 \times K_{sat}$ (m s^{-1})	E_t^a (%)	E_{50} (%)
Conventionally cultivated[b]	4.2	60	7
Direct-drilled	10.1	58	5
Direct-drilled, with worm channels plugged	2.2	ND	ND

Source: After Douglas *et al.* (1980/81); reprinted with permission of Elsevier Science Publishers.

ND = not determined.

[a]Calculated assuming the soil particle density was 2.6 g cm^{-3}.

[b]Cultivations included moldboard plowing to 25 cm depth.

macropores (earthworm channels) than direct-drilling ("no-till"), thus impairing the soils saturated hydraulic conductivity.

Macropores here are defined as voids or channels greater than 100 μm in equivalent diameter. The amount of macroporosity is considered as a volume fraction or percent of the total soil volume and will be designed by the symbol E_{100}.

Pores >100 μm equivalent diameter were chosen principally because root diameters for the majority of agricultural crops are greater than 100 μm (E.W. Russell, 1973; Gibbs, 1986). The presence of pores with diameters equal to or greater than the root diameter is crucial for the growth of crop roots (Goss, 1977). Pores with equivalent diameters in the range 100–1000 μm are often produced by crop roots (Tippkötter, 1983). Lance (1987) surveyed 45 silt loam sites with different management histories in Canterbury, New Zealand, and found E_{100} to be a versatile and meaningful index of topsoil structure. E_{100} was well correlated with other structural indices such as total porosity, dry bulk density, steady infiltration rate, and penetrometer resistance.

The present model is designed principally to apply to situations in modern mechanized agriculture. It is not designed for situations where fauna other than earthworms have an appreciable effect on soil macroposity or where significant freezing of the soil occurs.

Processes recognized in the model to create macropores are tillage, the action of root growth and decomposition, the activity of earthworms, and irreversible shrinkage of the soil. Processes that destroy macropores (generally the action of weather and in some cases tillage) are moderated by the level of macropore stability. In addition, macropores may be lost temporarily through blockage by roots. Soil water content is envisaged as being very important; it directly affects E_{100} via shrinkage and also controls biological activities, macropore collapse or blockage, and the ease and success of tillage operations.

It is intended that eventually the model can be used to describe changes in E_{100} throughout the soil profile. However, because of marked vertical variability in most of the processes described, at first it is necessary to confine application of the model to a particular region, say part of the A_p horizon, where vertical variability can be ignored to a reasonable approximation.

B. Model Assumptions

To keep the present model to a manageable size, a number of assumptions had to be made. Many minor assumptions are discussed in later sections, but the most major ones are listed here:

1. Crop nutrition remains satisfactory.
2. Soil temperature effects need not be included explicitly.
3. Compaction by animals and machines not involved with tillage operations is negligible.

All of these assumptions could probably be removed in subsequent developments of the model.

C. Model Structure

The model is summarized in Figure 2. It contains nine state variables, two of which are agencies external to the soil (tillage and weather). Six of the state variables have a direct influence on macroporosity (E_{100}). Most of the state variables (e.g., weather and soil organic matter) are in fact composite variables. Similarly time (t) is composite in the sense that for some processes it may be best expressed in days or even seconds (i.e., "seasonal time"), but for other processes developmental (or thermal) time may be more appropriate.

1. Crop Growth and Death

The term "crop canopy" as used here refers to both shoot dry matter and green area index (GAI). The latter is utilized as an indicator of the fractional amount of incident radiation intercepted by the crop (f_i):

$$f_i = 1 - \exp(-k \cdot \text{GAI}) \tag{1}$$

where k is an extinction coefficient. k may vary between crops, but is often in the range 0.3–0.5 (Monteith, 1965; Husain, 1984). GAI is used in preference to leaf area index so that the calculated values of f_i give a better indication of canopy protection of the soil surface (see section II.C.5).

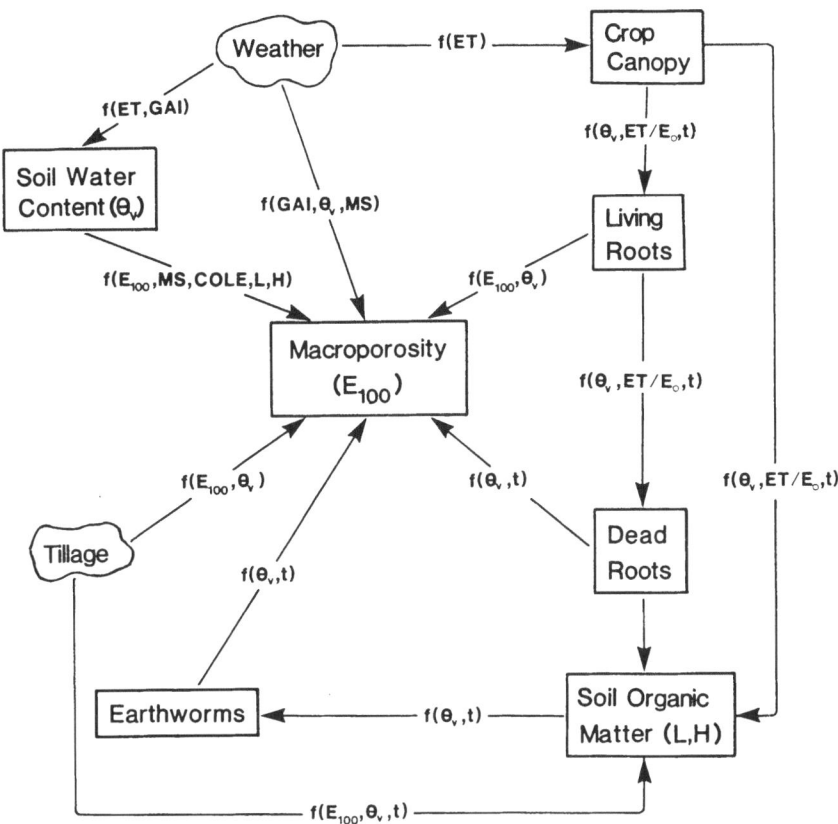

Figure 2. Conceptual model of changes in soil structure under different cropping systems. f = function; ET = actual evapotranspiration (m d^{-1}); E_o = potential evapotranspiration (m d^{-1}); GAI = green area index (m^2 m^{-2}); θ_v = soil water content (m^3 m^{-3}); COLE = coefficient of linear expansibility (m m^{-1}); E_{100} = pores >100 μm in diameter (m^3 m^{-3} or % v/v); MS = macropore stability; t = time (usually in days, see text); L = light or litter fraction (kg m^{-3}); H = heavy or humidified fraction (kg m^{-3}).

Shoot dry matter and GAI are assumed to be closely related, although the relationship will differ between crops and developmental phases.

The growth of the crop canopy is considered to be a simple function of actual evapotranspiration (ET) (Day et al., 1978; Hanks, 1983; Ritchie, 1983). ET is regarded as a function of potential evapotranspiration (E_o), soil water content (θ_v), and the amount and distribution of living roots. ET can be predicted quite readily using empirical relationships between ET, E_o, and θ_v if the maximum rooting depth is known (McAneney and Kerr, 1984). If the ration of ET/E_o becomes less than unity for several consecutive days, the shoot system may be considered to be under drought stress.

Such situations can be improved by rain, irrigation, or the crop sending new roots into soil with readily available water.

Growth of the root system is dependent on the size of the crop canopy and is modified by ET/E_o (as an index of drought stress), soil water content (θ_v), and time (t) (see, e.g., Huck and Hillel, 1983). In this pathway, t represents both developmental and seasonal time. The pathways of root and shoot death are also described as a function of ET/E_o, θ_v, and t (a simplification of the approach of Huck and Hillel, 1983). If the ratio of ET/E_o remains close to zero for too long, then the crop dies. The effect of water stress may depend on the developmental stage of the crop (Hanks, 1983), although there is some disagreement over this (Husain, 1984). Once again, t represents both developmental and seasonal time.

The concept of using the ratio ET/E_o as employed here has been much discussed in crop modelling work (Dougherty, 1976; Ritchie, 1981). However, it is difficult to find published crop growth models that take into account root and shoot death in a manner suitable for use here.

Clearly the root growth and death submodel proposed here needs further development and validation. Alternative models to describe root growth have been published. However, for the present purpose these have little advantage over the one proposed here, since they require either a great deal of information (e.g., Dexter, 1978a; Huck and Hillel, 1983) or are empirical (e.g., Gerwitz and Page, 1974). When applying the present model of changes in soil structure, the amount of living and dead roots should be measured for each situation being described (see section III).

The effects on crop growth of impaired root zone aeration are not considered at present.

2. Root Effects on Macroporosity

a. Living Roots

When extending, root must grow into existing pores larger than their own diameter, enlarge smaller pores, or create entirely new ones (Wiersum, 1957; Goss, 1977). Enlarging pores or creating new pores often necessitates the compaction of soil immediately around the root (Barley, 1954; Greacen et al., 1968) and may greatly reduce root extension rate (Goss, 1977). Roots appear to grow preferentially into macropores where these are accessible (R.S. Russell, 1977; Ehlers et al., 1983; Whiteley and Dexter, 1983). In the model developed here, living roots are therefore seen to change E_{100} not only by blocking existing macropores but also by creating and initially blocking new macropores. There is evidence to suggest that such processes can appreciably influence soil physical properties (Barley, 1953, 1954; Sedgley and Barley, 1958; Green and Fordham, 1975). However, Wiersum (1961) cautioned that characteristics that made a pore suitable for root penetration might not necessarily make that pore suitable for air and water transmission. In the model, the change in E_{100} caused by living roots is described as a function of θ_v, which influences penetration resistance into pores smaller in diameter than the roots (Barley and Greacen, 1967; Ehlers *et al.*, 1983), and the antecedent E_{100}, which determines the amount of existing pores wider than the roots and which can be freely penetrated.

There is a suggestion in the literature (Low, 1976) that root growth and soil shrinkage caused by water uptake may open up planes of weakness in aggregates. Shrinkage is dealt with in section II.C.5. If root growth *per se* opened up planes of weakness then E_{100} could be changed rather more than the volume of roots might suggest. However there is little evidence for such effects of root growth in situations for which the model has been designed.

b. Dead Roots

The present model suggests that a change in E_{100} can be expected when roots die and their channels are freed by root decomposition. There is then the possibility that such channels become available not only for root growth of a subsequent crop but also for movement of water through the soil (A.L. Clarke *et al.*, 1967; E.W. Russell, 1975). For example, Barley (1954) showed that saturated hydraulic conductivity of a sandy loam was significantly decreased by pore blockage as roots of maize grew, but was subsequently increased as the roots decomposed and root channels were freed for water movement.

Dead intact roots themselves have also been thought to contribute to water conduction (Emerson, 1954). Such effects may be small compared to the increased conductivity from decomposed roots (Barley and Sedgley, 1959).

The effects of root decomposition on E_{100} and the hydraulic properties of field soils are poorly documented. There is little published information concerning how long roots live in the field, how long it takes for dead roots to free macropores, and how significantly root channels affect water movement and subsequent root growth.

In the present model, the decomposition of a root is considered to be a function of θ_v and t. Dead roots are also considered to be a component of soil organic matter.

Decomposition processes are strongly influenced by changes in θ_v (F.E. Clarke, 1967; Doran, 1980; Orchard and Cook, 1983). Information concerning the time taken for dead roots to decompose is scarce, although decomposition rates for whole plant material may be predicted with reasonable confidence. For example, Jenkinson and Rayner (1977) measured a half-life of 0.17 years for decomposing plant material in the field. The results of Jenkinson (1977a) suggest a half-life of about 0.26 years for ryegrass (*Lolium perenne*) shoots and tops added to two soils in a laboratory; a similar figure can be estimated from his results for the same plant materials incubated in the same soils in the field. Some plant materials have a much longer half-life; figures varying from 2 to 8 years have been recorded under temperate climates (Shields and Paul, 1973; Jenkinson and Rayner, 1977). Half-lives for native soil organic matter appear to vary with soil clay content and type (Hart, 1984), although there is no evidence that the decomposition of crop roots is similarly affected *in vivo*.

3. Tillage

Tillage is any physical manipulation of a soil that changes its structure, strength, or position in order to improve conditions for crop production (Marshall and Holmes, 1979). The direct mechanical action of tillage affects the soil pore space and can thereby strongly influence soil transmission properties and root growth (Barber, 1971; Ehlers, 1977; Goss *et al.*, 1978; Ellis and Barnes, 1980; Ball, 1981; Klute, 1982; and see Tables 1 and 2). Tillage has also been shown to have other indirect effects on certain biological components of the soil (Rovira and Greacen, 1957; Barley, 1959; Low, 1972; Barnes and Ellis, 1979; Tisdall, 1985) that may in turn influence soil physical processes.

Here tillage is assumed to directly effect E_{100} by mechanical rearrangement of the soil. It is also assumed to indirectly affect E_{100} via its effects on soil organic matter, which may control earthworm population levels and the stability of macropores.

The direct effect of a given tillage operation on E_{100} is presented as a function of the antecedent values of E_{100} and θ_v. The results of Allmaras *et al.* (1967) support the use of antecedent E_{100} in this context, providing that, for their experiments, E_{100} was simply and closely related to total porosity. Unfortunately, Allmaras *et al.* (1967) do not quote values of E_{100}, but Gibbs (1986) and Lance (1987) have both observed simple, close relationships between E_{100} and total porosity in silt loams (Figure 3). Clearly, if the antecedent value of E_{100} is large, then tillage may not further increase it. Indeed some tillage operations (e.g., rolling) may *decrease* E_{100}. The size of θ_v strongly influences soil consistency and compressibility and therefore greatly affects the way in which tillage alters soil structure (Allmaras *et al.*, 1967; Archer, 1975; Boekel, 1979; Koolen and Kuipers, 1983). For example, tillage implements and tractor wheels can be expected to smear and compact a soil readily if it is wetter than its plastic limit, but if θ_v is below that limit, the soil may behave as a friable material (Marshall and Holmes, 1979). When θ_v in the hard-consistency range, tillage operations may produce clods and dust (Archer, 1975).

Tillage also causes losses of soil organic matter through increased microbial respiration. Rovira and Greacen (1957) reviewed the general mechanisms involved in soil organic matter loss under tillage and demonstrated in laboratory experiments that even mild tillage stimulated microbial activity. Three reasons were suggested for this: improved aeration, better distribu-

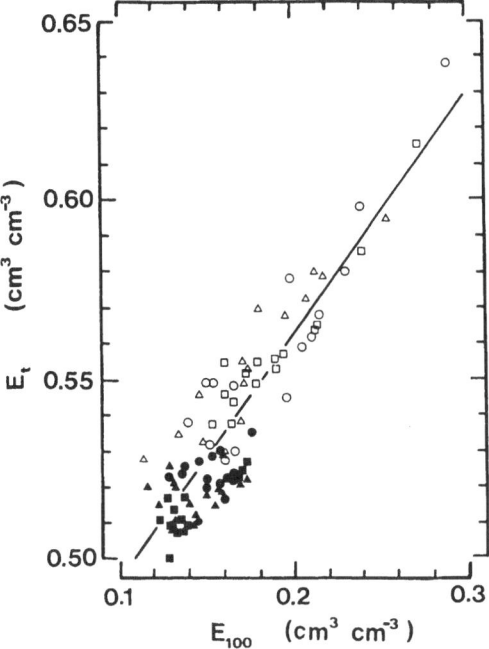

Figure 3. Relation between macroporosity (E_{100}) and total porosity (E_t) for a Wakanui silt loam (Aquic Dystric Utrochrept) at 5–15 cm depth (Gibbs, 1986). Measurements were made at intervals over 2 years of a field experiment with six different treatments. Open symbols represent soil conventionally cultivated (including plowing to 20 cm depth) annually. Solid symbols represent direct drilled soil. Circles represent land sown to winter wheat; triangles, land sown to perennial ryegrass; and squares, land kept free of plants. Each symbol represents the mean of four replicates. The line drawn is for the least-squares regression:

$$E_t = 0.426 + 0.68\ E_{100} \quad (r^2 = 0.82^{***})$$

tion of bacterial and fungal hyphae, and exposure of previously occluded organic matter to microbial attack. All of these changes can lead to a decrease in soil structural stability. Tisdall *et al.* (1978) reported that simulated tillage decreased the aggregate stability of a fine sandy loam. Their results suggested that severe restriction of microbial activity after tillage (e.g., if the soil remained dry) increased the net effect of the physical disruption.

Here it is suggested that the effect of tillage on soil organic matter can be described as a function of E_{100}, θ_v, and t, where t represents seasonal time. Soil water content has already been shown to be important in governing decomposition processes (see above). E_{100} is included as a factor here because the amount of macroporosity after tillage would strongly influence not only soil aeration (in conjunction with θ_v) but also the amount of pores of <100 μm effective diameter, some of which may be fine enough to occlude soil organic matter from microbial attack.

In general, liquid and plastic limits increase with soil organic matter content (Marshall and Holmes, 1979), and compactability decreases (Free et al., 1947; M.B. Russell et al., 1952). Thus, an appreciable change in the amount of soil organic matter may change the dependance on θ_v of the effects of tillage. Further detailed development of the model might require allowance for this effect, although such sophistication is not necessary for present purposes.

4. Earthworm Populations and Activities

The effects of earthworms on soil physical properties can largely be interpreted in terms of a change in pore size distribution (Syers and Springett, 1984). Earthworm burrowing may greatly influence total and macroporosity (Figure 4), saturated hydraulic conductivity (Table 2), root growth (R.S. Russell, 1977; Ehlers et al. 1983), and water infiltration (Ehlers, 1975; Scotter and Kanchanasut, 1981; Bouma et al., 1982; Springett, 1985). Little quantitative information is available on the construction and longev-

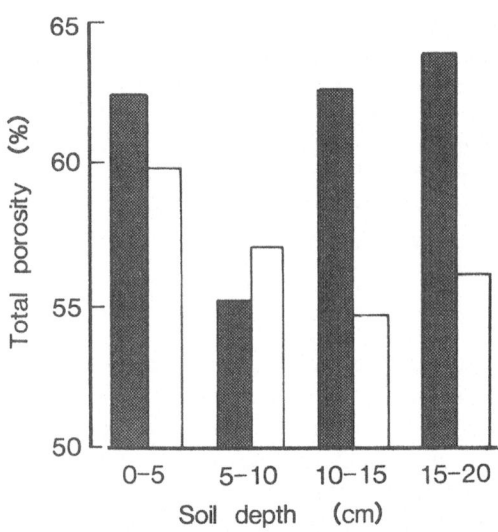

Figure 4. Effect on total porosity of introduction of the earthworm *Aporrectodea longa* (formerly *Allolobophora longa*). The soil was a Matapiro silt loam (Typic Durustalf) under pasture at Porongahau, New Zealand. The stippled bars are for areas where *A. longa* had been introduced 18 months previously. Open bars are for areas with resident worms (*A. caliginosa, A. trapezoides*) only (from Springett, 1985; reprinted with permission of Blackwell Scientific Publications).

ity of these burrows. Experiments involving earthworms have tended to concentrate on population dynamics (Ellis *et al.*, 1977; Barnes and Ellis, 1979; Lofs-Holmin, 1983; Edwards and Lofty, 1982; Clutterbuck and Hodgson, 1984) and rarely yield suitable quantitative information on burrowing. Besides those cited above, studies that have specifically examined the effects of earthworms on relevant soil physical properties include Evans (1984), Barley (1959), and Springett (1983).

Here it is suggested that changes in E_{100} caused by the burrowing of a given earthworm population can be described as a function of θ_v and t, where t represents both seasonal and developmental time.

Soil water content and seasonal time are both known to affect the aestivation characteristics and vertical distribution of earthworms (Barley, 1961; Edwards and Lofty, 1977). For example, Evans and Guild (1947, cited in Edwards and Lofty, 1977) showed that earthworm activity (as measured by the production of earthworm casts) followed a seasonal pattern and fluctuated with θ_v. Barley (1959) estimated that near Adelaide, Australia, earthworms were active for 24 weeks of the year, and he observed that they aestivated at 15–60 cm soil depth.

Soil water content may conceivably influence earthworm burrowing through its effect on soil strength. Dexter (1978b) reported that, over the range of 0.3–3 MPa, soil strength did not affect burrowing by *Aporroctodea caliginosa* (formerly *Allolobophora caliginosa*) through compressed soil samples at a matric potential of −100 cm. However, this result may not be general for soils at different water contents.

The pathway that changes E_{100} by earthworm burrowing might also include antecedent E_{100} and soil organic matter content as governing factors. Lance (1987) found that under laboratory conditions burrowing by *A. caliginosa* decreased with decreasing antecedent total porosity. Certain earthworm species are known to have permanent burrow systems (Edwards and Lofty, 1977), which would suggest that, if a satisfactory burrow system already existed and the food supply was adequate, little further burrowing might occur. Lance (personal communication, 1985) observed that in the laboratory, *A. caliginosa* may move preferentially through existing burrows rather than create new ones and has suggested that this behavior may be very dependent on the amount of available food in the soil. Although results of Gibbs (1986) offer indirect support for this notion, direct field evidence is scarce. Results from the laboratory experiments of Martin (1982) suggest that burrowing by *A. trapezoides* and *Lumbricus rubellus* decreases with increasing food supply. Martin's results also suggest that a proportion of the burrows and feeding cavities formed by species like *A. caliginosa*, *A. trapezoides*, and *L. rubellus* may be filled in or occluded by subsurface egestion under some conditions. The real pathway by which earthworms change E_{100} is undoubtedly more complex than the model suggests. However, the present simple approach may still prove adequate if detailed mathematical formulation of the model is attempted.

   Soil organic matter is often the single most important factor governing
the size of earthworm populations (Satchell, 1967; Lofs-Holmin, 1983).
Established grassland generally contains greater earthworm populations
than continuously tilled land (Barley, 1959; Edwards and Lofty, 1977), and
a decline in population may begin soon after grassland is ploughed up
(Low, 1972; E.W. Russell, 1973). Although a shortage of soil organic matter is considered largely responsible for decreased earthworm populations
in arable soils compared to grassland (Edwards and Lofty, 1982), lack of a
crop protecting the soil surface from drying may also contribute (Lofs-
Holmin, 1983). Both effects are amenable to analysis using this model.

   In the model, the effect of soil organic matter on earthworm populations
is suggested to depend on θ_v and t (Figure 2). Here again, t may represent
developmental as well as seasonal time, because the state of earthworm
maturity is important to both population dynamics and activities. The
importance of θ_v to earthworm populations is suggested indirectly by the
observed effects of rainfall (Barnes and Ellis, 1979; McColl, 1984).

   Finally, no distinction has been made at this stage between different
earthworm species, although these may have different burrowing habits
(Edwards and Lofty, 1977; Springett, 1983). It is difficult to see how the
activities of different species can be distinguished in the model until more
detailed, quantitative information on earthworm burrowing becomes available.

5. Weather Effects

The weather parameters considered most appropriate for the model are
rainfall and potential evapotranspiration (E_o). For convenience, rainfall
and irrigation are treated identically here. E_o is a function of air temperature, vapor pressure deficit, net radiation, and crop cover (Ritchie, 1972).
In the model, weather not only has a direct influence on E_{100} but also has
an indirect influence via θ_v and the effects of weather on crop growth.

   The direct effects of weather on E_{100} considered here are water-drop
impact, slaking, and dispersion. A common cause of pore blockage near
the soil surface is dispersed or slaked material being washed into the soil
profile under intense rainfall or irrigation (McIntyre, 1958; Awadhal and
Thierstein, 1985). Deeper in the profile, slaking and dispersion of particularly weakly structured horizons may also cause loss of porosity (Hillel,
1980).

   In the model, the direct effects E_{100} of a given rainfall or irrigation event
are assumed to depend on macropore stability, the initial θ_v, and GAI.
Cernuda et al. (1954) found that in several Puerto Rican soils, decreasing
the antecedent θ_v increased the ease with which natural soil surface aggregates were disrupted by water drops. The work of Barley (1953) and Ojeniyi
and Dexter (1983) suggests indirectly that crop cover can reduce the potential
for aggregate disruption by direct impact of raindrops or irrigation water.

More direct evidence for this comes from studies of soil erosion (Fullen, 1985). For simplicity, at this stage, the fractional area of the soil surface protected by the crop canopy from rain is assumed to aproximate the fraction of incident radiation intercepted by the crop (f_i), which is dictated by GAI (Eq. 1). Variations in rainfall intensity are not considered in the present model.

The indirect effects of weather on E_{100} considered mostly concern slumping and irreversible shrinkage due to soil drying. Slumping is defined here as the compaction of soil under its own weight. In some field soils, pore blockage and collapse may be particularly pronounced when rainfall follows tillage (Baker, 1979; Hamblin, 1982; Dexter et al., 1983; Ojeniyi and Dexter, 1983). However, we are unaware of instances in which such effects can be directly attributed to slumping, rather than slaking alone or in combination with slumping. It is difficult to say how significant slumping is in the context of the whole model, and research is clearly needed. We suggest that slumping occurs when θ_v is large and its extent is moderated by macropore stability and the antecedent E_{100} (with a large antecedent E_{100} increasing the likely amount of slumping).

Although the model does not account for reversible swelling and shrinking, changes in E_{100} due to irreversible shrinkage are included. Even in soils without large clay contents, drying might cause significant shrinkage (Stirk, 1954). Drying processes may increase the adsorption of aggregate stabilizing agents onto mineral surfaces (Reid and Goss, 1982a), which may effectively make the associated shrinkage largely irreversible.

At present, it is difficult to predict the effects of irreversible shrinkage on E_{100} of anisotropic field soils. In general, irreversible shrinkage will increase E_{100}, since the bulk soil volume loss often exceeds the loss of soil volume due to the reduction in soil surface height (Newman and Thomasson, 1979; Towner, 1986). Experimental evidence of wetting and drying cycles affecting E_{100} is scant. Newman and Thomasson (1979) showed an increase in the volume of pores >29 μm effective diameter in a clay topsoil after a dry season, but generally such pores seemed little affected by drying in the other five soils studied. However, the small size of the samples measured may have precluded detection of significant effects of soil shrinkage upon macroporosity. Sartori et al. (1985) demonstrated that 6–16 wetting and drying cycles more than doubled the volume of pores 30–1000 μm effective diameter in a puddled, saline, clay soil (Figure 5).

The maximum irreversible increase in E_{100} that drying could cause is strongly influenced by the soil's coefficient of linear extensibility (COLE). Although COLE may be a function of θ_v (Stirk, 1954), it is difficult to appraise its sensitivity to other parameters or processes in this model. We are not aware of any evidence that COLE is directly affected by tillage, roots, or earthworm activities. Increasing soil organic matter has been reported to increase COLE (Reeve et al., 1980), decrease it (Davidson and Page, 1956), and have no measurable effect (Franzmeier and Ross, 1968).

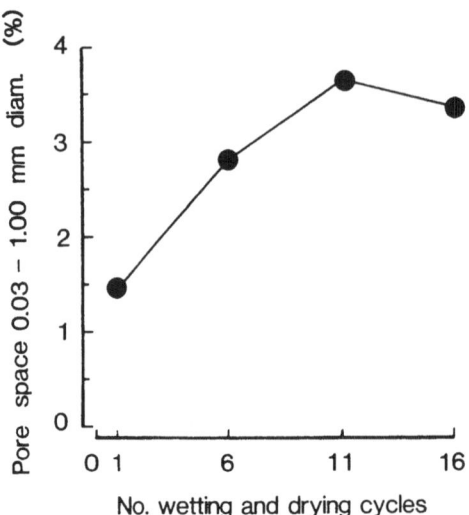

Figure 5. Influence of wetting and drying cycles on the volume of pores 0.03–1.00 mm equivalent diameter (as a % of total soil volume). The soil was a puddled, saline clay (Xerollic Camborthid), and the soils were cycled between saturation and air dry in a laboratory (Sartori *et al.*, 1985; reprinted by permission of Williams & Wilkins).

Neither Reeve *et al.* (1980) nor Franzmeier and Ross (1968) appeared to estimate COLE in the high range of soil water contents where "structural" rather than "normal" or "residual" shrinkage (Stirk, 1954) may occur. "Structural" shrinkage is possibly the most important shrinkage phase for this model, and has been shown to be sensitive to changes in organic matter levels (Davidson and Page, 1956) and cropping system (Chan, 1982). The size of COLE during each shrinkage phase, and the soil water content ranges for each phase, determine the maximum possible net effect of shrinkage on E_{100} (*net* because intraaggregate macroporosity will be decreased by shrinkage whereas interaggregate macroporosity is often increased).

In the present model, we suggest that the effect of irreversible soil shrinkage upon E_{100} depends on θ_v and COLE (which influence the volume change as the soil dries) and the L and H fractions of the organic matter (which influence the extent to which the shrunken soil structure is stabilized and resistant to reswelling). Research is clearly needed to test this hypothesis and to elucidate the effects of soil organic matter on COLE.

6. Soil Organic Matter

Soil organic matter has so far been discussed as a single substance rather than as a highly complex material. The model recognizes only two soil organic matter fractions—the light, or litter, fraction (L), and the heavy, or humified, fraction (H). Crop residues are assumed to contribute directly to the light fraction. For simplicity, all the products of decomposition are termed heavy. Conversion of the light material into heavy material is probably dependent mostly on seasonal time, soil water content, and the activities of living roots (Jenkinson and Rayner, 1977; Jenkinson, 1977b; Orchard and Cook, 1983; Reid and Goss, 1982b, 1983; Sparling et al., 1982). Here again, soil temperature effects are not dealt with explicity in the model. For the sake of simplicity, the present model does not cover situations where significant applications of organic mulches or manures are made.

It should be realized that the soil organic matter state variable is composite in nature and the result of many interacting processes. The role of soil organic matter per se has therefore been greatly simplified to fit into the overall model structure. Both the light and heavy fractions in the model act as energy sources for earthworms and influence macropore stability. Dead roots are parts of the light fraction and, as already mentioned, directly influence E_{100}.

7. Macropore Stability

Macropore stability (MS) was introduced to represent the ability of pores >100 μm effective diameter to withstand collapse under the action of water. Unfortunately, there are no commonly used techniques for measuring MS of field soils, either in situ or as "undisturbed" core samples. Consequently, current knowledge of the sensitivity of MS to different factors is poor. Even when porosity measurements are made, the stability of the pore system is often neglected (Sequi, 1978).

However, there is much indirect evidence that MS can be influenced especially by (1) the activities of living roots, earthworms, and microorganisms; (2) tillage and subsequent rheolitic processes; (3) the soil content of organic matter (particularly polysaccharides), iron, and aluminium; and (4) climatic effects such as wetting and drying or freezing and thawing cycles (Robinson and Jacques, 1958; Harris et al., 1966; Blake and Gilman, 1970; E.W. Russell, 1971, 1973; Arya and Blake, 1972; Richardson, 1976; Giovannini and Sequi, 1976; Hamblin and Greenland, 1977; Edwards and Lofty, 1977; Quirk, 1978; Tisdall and Oades, 1979; Reid and Goss, 1981, 1982a; Utomo and Dexter, 1981; Reid et al., 1982).

This indirect evidence come chiefly from studies of aggregate stability, making the assumption that aggregate stability and macropore stability will be similarly affected by these factors. Indeed, the intense attention focused on aggregate stability in earlier work had its foundation in the assumption that pores between aggregates would only persist when the aggregates had considerable stability (Greenland, 1977). Despite this, even now it is difficult to suggest ways in which aggregate stability results can be related directly and quantitatively to the stability of the pore regime, especially when aggregate stability is measured by wet-sieving (Beacher and Strickling, 1955; Bruce, 1955).

Further development of the model might require that either links between MS and aggregate stability measurements be identified, or that new evidence be obtained to define how biological and physicochemical factors influence MS. However, a satisfactory means to determine MS for field soils must be developed first.

We suggest that the above uncertainties associated with MS do not prevent the model in its present form from having useful applications.

III. General Discussion

The conceptual model described above is an attempt to systematically summarize knowledge of the major processes known to create, destroy, or block macropores. Certainly the model represents a greatly simplified approach to a complex problem, and it does not include all of the possible factors and interactions between them. However, we suggest that it includes the most important processes relevant to modern mechanized farming. Further refinements and extension are to be encouraged, but we consider that the model should be tested practically before much more is added. Such testing is already under way in New Zealand, and the results of validation exercises are being prepared for publication.

Derivation of this model has revealed many important areas of uncertainty in the literature. In particular, there is a need to know much more about root longevity, earthworm burrowing, the effects of soil shrinkage on E_{100}, and the factors governing MS. There is a real need for a simple and realistic method to measure MS directly on undisturbed soil samples.

Despite the model's present simplicity, it may still be used effectively. For instance, it can be used to investigate the relative importance of the processes known to change macroporosity in a particular part of a soil profile. A form of budget equation can be devised from the model:

$$\Delta E_{100} = T + B_e + R_d - R_g - S \tag{2}$$

where ΔE_{100} is the net change in macroporosity over a given period, T is the change due to tillage, B_e is the change due to earthworm burrowing, R_d is the change due to decomposition of roots, R_g is the change due to root growth, and S is the change in macroporosity due to slaking and slumping and/or irreversible shrinkage.

To apply Eq. 2, ΔE_{100} could be estimated from measurements of E_{100}, perhaps made using the common water retention method of Leamer and Lutz (1940), at the start and end of a study period. T could be estimated by measuring E_{100} immediately before and after tillage. B_e could be estimated perhaps from observations of worm channel numbers, diameters, and lengths in the samples used to measure E_{100}. R_d and R_g would be more difficult to estimate, since it is difficult to know the proportion of roots likely to create new macropores rather than grow into existing ones. Nevertheless, robust best and worst case estimates could still be made if the volumes of living and dead roots per unit soil volume were measured. Finally, S would have to be inferred from Eq. 2 once estimates of the other parameters have been entered.

For maximum accuracy when characterizing the effect of a cropping system, Eq. 2 might need to be solved for several short consecutive periods, especially if tillage is intermittent and if S can be expected to be rapid. Identifying the slaking, slumping, and shrinkage contributions to S would be difficult for many soils at present. Irreversible shrinkage might be reasonably easily estimated if E_{100} is measured at the start and end of each substantial drying phase. In many climates, this would necessitate frequent measurements. Where such frequency is impractical, however, it may still be possible to interpret values of S quite simply. For instance, in well-drained soils that are particularly resistant to slaking (e.g., Emerson (1967) classes 1 and 2), slaking can be discounted except perhaps at the very soil surface, and for horizons close to the surface slumping may be expected to be negligible, leaving irreversible shrinkage as the dominant contributor to S.

Used in the manner described above, the model can assist in experiment design, particularly in the choice of measurements and measurement frequency. Furthermore, such relatively simple calculations based on the model would start to reveal in a quantitative way the relative importance of different processes that change E_{100} under different cropping systems. The results of such calculations could be extremely useful when attempting to improve soil management strategies.

IV. Summary

Soil structure can be markedly affected by the choice of cropping system. Such changes are often very important because of their strong influence on crop growth, the potential for soil erosion, and the ease of land and crop management (especially under wet conditions).

Despite the large number of experimental investigations, it remains difficult to predict, or even interpret, changes in soil structure under different cropping systems. This difficulty arises because of the large number of possible combinations of soil type, crop, management and climate, and because most field experiments have yielded site-, season-, and technique-specific results. To help solve these problems this paper presents a conceptual model to assist the design and interpretation of experiments concerned with effects of cropping systems on soil structure. The model also has potential for mathematical development for predictive purposes.

Soil pores > 100 μm in effective diameter are of great importance for root growth and soil transmission properties. Therefore the model is concentrated on the creation and destruction of such pores. Processes recognized to create pores in this size range are tillage, root growth and decomposition, earthworm activity, and irreversible shrinkage of the soil. These pores are envisaged as being destroyed by slaking and slumping of the soil under the action of rain or irrigation, and they may be temporarily blocked by root growth. Factors involved in the processes of macropore creation, destruction, and blockage are reviewed. The model does not explicitly account for crop nutrition, soil temperature, or compaction by animals and machines not involved with tillage operations. Development of the model has indicated several areas where further research is needed. In particular, research is needed on root longevity, earthworm burrowing, the effects of soil shrinkage on macropores, and the characterization of macropore stability.

One method of applying the model is outlined. This method uses a form of budget equation to assess the relative importance of several factors known to influence changes in soil structure under cropping management. Such applications may be useful when deciding practical strategies to manage soil structure.

Acknowledgments

R.J. Gibbs acknowledges the sponsorship of the New Zealand Government under the Commonwealth Scholarship Scheme. We also thank the Lincoln College Research Committee for supporting this work, and C.D. Lance, J.A. Springett, and G.J. Churchman for helpful discussions.

References

Ahuja, L.R., J.W. Naney, R.E. Green, and D.R. Nielsen. 1984. Macroporosity to characterise spatial variability of hydraulic conductivity and effects of land management. *Soil Sci. Soc. Am. J.* 48:69–72.

Allmaras, R.R., R.E. Burwell, and R.F. Holt. 1967. Plow-layer porosity and surface roughness from tillage as affected by initial porosity and soil moisture at tillage time. *Soil Sci. Soc. Am. Proc.* 31:550–556.

Allmaras, R.R., K. Ward, C.L. Douglas, and L.G. Ekin. 1982. Long term cultivation effects on hydraulic properties of a Walla-Walla silt loam. *Soil Tillage Res.* 2:265–279.

Archer, J.R. 1975. Soil consistency. In: *Soil Physical Conditions and Crop Production. MAFF Tech. Bull.* 29:289–297, HMSO, London.

Arya, L.M., and G.R. Blake. 1972. Stabilization of newly formed aggregates. *Agron. J.* 64:177–180.

Awadhal, N.K., and G.E. Thierstein. 1985. Soil crust and its impact on crop establishment: A review. *Soil Tillage Res.* 5:289–302.

Baker, S.W. 1979. Pore size distribution—a factor to be considered in infiltration studies. *J. Hydrol.* 41:279–290.

Ball, B.C. 1981. Pore characteristics of soils from two cultivation experiments as shown by gas diffusivities and permeabilities and air-filled porosities. *J. Soil Sci.* 32:483–498.

Barber, S.A. 1971. Effect of tillage on corn (*Zea mays* L.) root distribution and morphology. *Agron. J.* 63:724–726.

Barley, K.P. 1953. The root growth of irrigated perennial pastures and its effect on soil structure. *Aust. J. Agric. Res.* 4:283–291.

Barley, K.P. 1954. Effects of root growth and decay on the permeability of a synthetic sandy loam. *Soil Sci.* 78:205–210.

Barley, K.P. 1959. The influence of earthworms on soil fertility. I. Earthworm populations found in agricultural land near Adelaide. *Aust. J. Agric. Res.* 10:171–178.

Barley, K.P. 1961. The abundance of earthworms in agricultural land and their possible significance in agriculture. *Adv. Agron.* 13:249–268.

Barley, K.P., and E.L. Greacen. 1967. Mechanical resistance as a soil factor influencing the growth of roots and underground shoots. *Adv. Agron.* 19:1–43.

Barley, K.P., and R.H. Sedgley. 1959. The influence of grass roots on the conduction of water through coarse-textured soils. *Soils Fertil.* 22:155–156.

Barnes, B.T., and F.B. Ellis. 1979. Effects of different methods of cultivation and direct drilling, and disposal of straw residues, on populations of earthworms. *J. Soil Sci.* 30:669–679.

Beacher, B.F., and E. Strickling. 1955. Effect of puddling on water stability and bulk density of aggregates of certain Maryland soils. *Soil Sci.* 80:363–373.

Blake, G.R. and R.D. Gilman. 1970. Thixotropic changes with ageing of synthetic soil aggregates. *Soil Sci. Soc. Am. Proc.* 34:561–564.

Boekel, P. 1979. The workability of the soil in spring in relation to moisture content and moisture transport. *Proc. Int. Soil Tillage Res. Organ. Conf.* 8:293–298.

Bouma, J., C.F.M. Belmans, and L.W. Dekker. 1982. Water infiltration and redistribution in a silt loam subsoil with vertical worm channels. *Soil Sci. Soc. Am. J.* 46:917–921.

Bridge, B.J., J.J. Mott, W.H. Winter, and R.J. Hartigan. 1983. Improvement in soil structure resulting from sown pastures on degraded areas in the dry savanna woodland of northern Australia. *Aust. J. Soil Res.* 21:83–90.

Bruce, R.R. 1955. An instrument for the determination of soil compactibility. *Soil Sci. Soc. Am. Proc.* 19:253–257.

Cannell, R.Q. 1985. Reduced tillage in north-west Europe—a review. *Soil Tillage Res.* 5:129–177.

Cary, J.W., and C.W. Hayden. 1973. An index for soil pore size distribution. *Geoderma* 9:249–256.

Cernuda, C.F., R.M. Smith, and J. Vicente-Chandler. 1954. Influence of initial soil moisture condition on resistance of macroaggregates to slaking and to waterdrop impact. *Soil Sci.* 77:19–27.

Chan, K.Y. 1982. Shrinkage characteristics of soil clods from a grey clay under intensive cultivation. *Aust. J. Soil Res.* 20:65–68.

Clarke, A.L., D.J. Greenland, and J.P. Quirk. 1967. Changes in some physical properties of the surface of an impoverished red-brown earth under pasture. *Aust. J. Soil Res.* 5:59–68.

Clarke, A.L., and J.S. Russell. 1977. The effect of sequential practices on soil physical properties. In: J.S. Russell and E.L. Greacen (eds). *Soil Factors in Crop Production in a Semi-Arid Environment.* University of Queensland Press, St. Lucia, Australia, pp. 279–300.

Clarke, F.E. 1967. Bacteria in soil. In: A. Burges and F. Raw (eds.) *Soil Biology.* Academic Press, London, pp. 15–49.

Clutterbuck, B.J., and D.R. Hodgson. 1984. Direct drilling and shallow cultivation compared with ploughing for spring barley on a clay loam in northern England. *J. Agric. Sci. (Camb.)* 102:127–134.

Cooke, G.W., and R.J.B. Williams. 1972. Problems with cultivation and soil structure at Saxmundham. *Roth. Rep. 1971* 2:122–142.

Coote, D.R., and J.F. Ramsey. 1983. Quantification of the effects of over 35 years of intensive cultivation on four soils. *Can. J. Soil Sci.* 63:1–14.

Davidson, S.E., and J.B. Page. 1956. Factors influencing swelling and shrinking in soils. *Soil Sci. Soc. Am. Proc.* 20:320–324.

Davies, D.B., D.J. Eagle, and J.B. Finney. 1982. *Soil Management,* 4th Ed. Farming Press, Ipswich, U.K.

Day, W., B.J. Legg, B.K. French, A.E. Johnston, D.W. Lawlor, and W.D. Jeffers. 1978. A drought experiment using mobile shelters: The effect of drought on barley yield, water use and nutrient uptake. *J. Agric. Sci. (Camb.)* 91:599–623.

Dexter, A.R. 1978a. A stochastic model for the growth of roots in tilled soil. *J. Soil Sci.* 29:102–116.

Dexter, A.R. 1978b. Tunnelling in soil by earthworms. *Soil Biol. Biochem.* 10:447–449.

Dexter, A.R., J.K. Radke, and J.S. Hewitt. 1983. Structure of a tilled soil as influenced by tillage, wheat cropping, and rainfall. *Soil Sci. Soc. Am. J.* 47:570–575.

Doran, J.W. 1980. Soil microbial and biochemical changes associated with reduced tillage. *Soil Sci. Soc. Am. J.* 44:765–771.

Dougherty, C.T. 1976. Water in the crop model SIMED 2. In: *Proceedings of Soil and Plant Water Symposium. Information Series. DSIR* 126:103–110.

Douglas, J.T., and M.J. Goss. 1982. Stability and organic matter content of surface soil aggregates under different methods of cultivation and in grassland. *Soil Tillage Res.* 2:155–175.

Douglas, J.T., M.J. Goss, and D. Hill. 1980/81. Measurements of pore characteristics in a clay soil under ploughing and direct drilling, including use of a radioactive (^{144}Ce) technique. *Soil Tillage Res.* 1:11–18.

Doyle, J.J., and F.G. Hamlyn. 1960. Effects of different cropping systems and of a soil conditioner (VAMA) on some soil physical properties and on growth of tomatoes. *Can. J. Soil Sci.* 40:89–98.

Edwards, C.A., and J.R. Lofty. 1977. *Biology of Earthworms,* 2d Ed. Chapman and Hall, London.

Edwards, C.A., and J.R. Lofty. 1982. The effect of direct drilling and minimal cultivation on earthworm populations. *J. Appl. Ecol.* 19:723–734.

Ehlers, W. 1975. Observations on earthworm channels and infiltration on tilled and untilled loess soil. *Soil Sci.* 119:242–249.

Ehlers, W. 1977. Measurement and calculation of hydraulic conductivity in horizons of tilled and untilled loess-derived soil, Germany. *Geoderma* 19:293–306.

Ehlers, W., U. Kopke, F. Hesse, and W. Bohm. 1983. Penetration resistance and root growth of oats in tilled and untilled loess soil. *Soil Tillage Res.* 3:261–275.

Ellis, F.B., and B.T. Barnes. 1980. Growth and development of root systems of winter cereals grown after different tillage methods including direct drilling. *Plant Soil* 55:283–295.

Ellis, F.B., J.B. Elliot, B.T. Barnes, and K.R. Howse. 1977. Comparison of direct drilling, reduced cultivation and ploughing on the growth of cereals. *J. Agric. Sci. (Camb.)* 89:631–642.

Emerson, W.W. 1954. Water conduction by severed grass roots. *J. Agric. Sci. (Camb.)* 45:241–245.

Emerson, W.W. 1967. A classification of soil aggregates based on their coherence in water. *Aust. J. Soil Res.* 5:47–57.

Evans, A.C. 1948. Studies on the relationship between earthworms and soil fertility. II. Some effects of earthworms on soil structure. *Ann. Appl. Biol.* 35: 1–13.

Franzmeier, D.P., and S.J. Ross. 1968. Soil swelling: Laboratory measurement and relation to other soil properties. *Soil Sci. Soc. Am. Proc.* 32:573–577.

Free. G.R., J. Lamb, and E.A. Carleton. 1947. Compactibility of certain soils as related to organic matter and erosion. *J. Am. Soc. Agron.* 39:1068–1076.

Fullen, M.A. 1985. Erosion of arable soils in Britain. *Int. J. Environ. Stud.* 26:55–69.

Germann, P., and K. Bevan. 1981. Water flow in soil macropores. I. An experimental approach. *J. Soil Sci.* 32:1–13.

Gerwitz, A., and E.R. Page. 1974. An empirical mathematical model to describe plant root systems. *J. Appl. Ecol.* 11:773–782.

Gibbs, R.J. 1986. Changes in soil structure under different cropping systems. Ph.D. Thesis, Lincoln College, University of Canterbury, Canterbury, New Zealand.

Giovannini, G., and P. Sequi. 1976. Iron and aluminium as cementing substances of soil aggregates. II. Changes in stability of soil aggregates following extraction of iron and alluminium by acetylactone in a non-polar solvent. *J. Soil Sci.* 27:148–153.

Goss, M.J. 1977. Effects of mechanical impedance on root growth in barley (*Hordeum vulgare* L.). I. Effects on elongation and branching of seminal roots. *J. Exp. Bot.* 28:96–111.

Goss, M.J., W. Ehlers, F.R. Boone, I. White, and K.R. Howse. 1984. Effects of soil management practice on soil physical conditions affecting root growth. *J. Agric. Eng. Res.* 30:131–140.

Goss, M.J., K.R. Howse, and W. Harris. 1978. Effects of cultivation on soil water retention and water use by cereals in clay soils. *J. Soil Sci.* 29:475–496.

Goss, M.J., and J.B. Reid. 1981. Interaction between crop roots and soil structure. In: J. Davies and F.E. Shotton, (eds.) *Aspects of Crop Growth*, MAFF Ref. Book 341. HMSO, London, pp. 34–48.

Greacen, E.L., D.A. Farrell, and B. Cockroft. 1968. Soil resistance to metal probes and plant roots. *Trans. Int. Cong. Soil Sci.* 9:769–779.

Green, R.D., and S.J. Fordham. 1975. A field method of measuring air permeability in soil. In: *Soil Physical Conditions and Crop Production*, MAFF Tech. Bull. HMSO, London, pp. 273–288.

Greenland, D.J. 1977. Soil damage by intensive arable cultivation: Temporary or permanent. *Phil. Trans. R. Soc. Lond. B* 281:193–208.

Hamblin, A.P. 1982. Soil water behavior in response to changes in soil structure. *J. Soil Sci.* 33:375–386.

Hamblin, A.P., and D.J. Greenland. 1977. Effect of organic constituents and complexed metal ions on aggregate stability of some East Anglian soils. *J. Soil Sci.* 28:410–416.

Hanks, R.J. 1983. Yield and water-use relationships: An overview. In: H.M. Taylor, W.R. Jordan, and T.R. Sinclair (eds.) *Limitations to Efficient Water Use in Crop Production.* American Society of Agronomy, Madison, WI, pp. 392–411.

Harris, R.F., G. Chesters, and O.N. Allen. 1966. Dynamics of soil aggregation. *Adv. Agron.* 18:107–169.

Hart, P.B.S. 1984. Effects of soil type and past cropping on the nitrogen supplying ability of arable soils. Ph.D. Thesis, University of Reading, Reading, U.K.

Hillel, D. 1980. *Applications of Soil Physics.* Academic Press, New York.

Huck, M.G., and D. Hillel. 1983. A model of root growth and water uptake accounting for photosynthesis, respiration, transpiration an soil hydraulics. In: D. Hillel (ed.) *Advances in Irrigation.* Academic Press, New York, pp. 272–333.

Huṣain, M.M. 1984. *The Response of Field Bean* (Vicia faba *L.*) *to Irrigation and Sowing Date.* Ph.D Thesis, Lincoln College, University of Canterbury, Canterbury, New Zealand.

Jenkinson, D.S. 1977a. Studies on the decomposition of plant material in soil. IV. The effect of rate of addition. *J. Soil Sci.* 28:417–423.

Jenkinson, D.S. 1977b. Studies on the decomposition of plant material in soil. V. The effects of plant cover and soil type on the loss of carbon from ^{14}C labelled ryegrass decomposing under field conditions. *J. Soil Sci.* 28:424–434.

Jenkinson, D.S., and J.H. Rayner. 1977. The turnover of soil organic matter in some of the Rothamstead classical experiments. *Soil Sci.* 123:298–305.

Klute, A. 1982. Tillage effects on the hydraulic properties of soil: A review. In: P.W. Unger and D.M. Van Doren (eds.) *Predicting Tillage Effects on Soil Physical Properties and Processes.* Spec. Pub. 44. American Society for Agronomy, Madison, WI, pp. 29–43.

Koolen, A.J., and H. Kuipers. 1983. *Agricultural Soil Mechanics.* Springer-Verlag. Berlin.

Lance, C.D. 1987. Changes in soil structure under various cropping systems. Ph.D. Thesis, Lincoln College, University of Canterbury, Canterbury, New Zealand.

Leamer, R.W., and J.F. Lutz. 1940. Determination of pore size distribution in soils. *Soil Sci.* 49:374–360.

Lofs-Holmin, A. 1983. Earthworm population dynamics in different agricultural rotations. In: J.E. Satchell (ed.) *Earthworm Ecology from Darwin to Vermiculture.* Chapman and Hall, London, pp. 151–170.

Low, A.J. 1972. The effect of cultivation on the structure and other physical characteristics of grassland and arable soils (1945–1970). *J. Soil Sci.* 23:363–380.

Low, A.J. 1976. Effects of long periods under grass on soils under British conditions. *J. Sci. Food Agric.* 27:571–582.

Marshall, T.J., and J.W. Holmes. 1979. *Soil Physics.* Cambridge University Press, London.

Martin, N.A. 1982. The interaction between organic matter in soil and the burrowing activity of three species of earthworms (Oligochaeta: Lumbricidae). Pedobiologia 24:185–190.

McAneney, K.J., and J.P. Kerr (eds.). 1984. *Environmental Inputs to Agronomic*

Research Guidelines. Agricultural Research Division, MAF, Wellington, New Zealand.

McColl, H.P. 1984. Nematicides and field population of enchytraeids and earthworms. *Soil Biol. Biochem.* 16:139–143.

McIntyre, D.S. 1958. Permeability of soil crusts formed by raindrop impact. *Soil Sci.* 85:185–189.

Monteith, J.L. 1965. Light distribution and photosynthesis in field crops. *Ann. Bot.* 29:17–37.

Newman, A.C.D., and A.J. Thomason. 1979. Rothamsted studies of soil structure. III. Pore size distributions and shrinkage processes. *J. Soil Sci.* 30:415–439.

Ojeniyi, S.O., and A.R. Dexter, 1983. Changes in the structure of differently tilled soil in a growing season. *Soil Tillage Res.* 3:39–46.

Orchard, V.A., and F.J. Cook. 1983. Relationship between soil respiration and soil moisture. *Soil Biol. Biochem.* 15:447–453.

Page, J.B., and C.J. Willard. 1946. Cropping systems and soil properties. *Soil Sci. Soc. Am. Proc.* 11:81–88.

Quirk, J.P. 1978. Some physico-chemical aspects of soil structural stability—a review. In: W.W. Emerson, R.D. Bond, and A.R. Dexter (eds.) *Modification of Soil Structure.* John Wiley, Chichester, U.K. pp. 3–16.

Reeve, M.J., D.G.M. Hall, and P. Bullock. 1980. The effect of soil composition and environmental factors on the shrinkage of some clayey British soils. *J. Soil Sci.* 31:429–442.

Reid, J.B., and M.J. Goss. 1981. Effect of living roots of different plant species on the aggregate stability of two arable soils. *J. Soil Sci.* 32:521–541.

Reid, J.B., and M.J. Goss, 1982a. Interactions between soil drying due to plant water use and decreases in aggregate stability caused by maize roots. *J. Soil Sci.* 33:47–53.

Reid, J.B., and M.J. Goss. 1982b. Suppression of decomposition of ^{14}C-labelled plant roots in the presenct of living roots of maize and perennial ryegrass. *J. Soil Sci.* 33:387–395.

Reid, J.B., and M.J. Goss. 1983. Growing crops and transformations of ^{14}C-labelled soil organic matter. *Soil Biol. Biochem.* 15:687–691.

Reid, J.B., M.J. Goss, and P.D. Robertson. 1982. Relationship between the decreases in soil stability effected by the growth of maize roots and changes in organically bound iron and aluminium *J. Soil Sci.* 33:397–410.

Richardson, S.J. 1976. Effect of artificial weathering cycles on the structural stability of a dispersed silt soil. *J. Soil Sci.* 27:287–294.

Ritchie, J.T. 1972. Model for predicting evaporation from a row crop with incomplete cover. *Water Resour. Res.* 8:1204–1213.

Ritchie, J.T. 1981. Water dynamics in the soil-plant-atmosphere system. *Plant Soil* 58:81–96.

Ritchie, J.T. 1983. Efficient water use in crop production: Discussion on the generality of relationships between biomass production and evapotranspiration. In: H.M. Taylor, W.R. Jordan, and T.R. Sinclair (eds.) *Limitations to Efficient Water Use in Crop Production.* American Society for Agronomy, Madison, WI, pp. 29–44.

Robinson, G.S., and W.A. Jacques. 1958. Root development in some common New Zealand pastures. X. Effect of pure sowings of some grasses and clovers on the structure of a Tokomaru silt loam. *N.Z. J. Agric. Res.* 28:209–219.

Ross, C.W., and K.A. Hughes. 1985. Maize/oats forage rotation under three cultivation practices, 1978–83. II. Soil properties. N.Z. J. Agric Res. 28:209–219.

Rovira, A.D., and E.L. Greacen. 1957. The effect of aggregate disruption on the activity of microorganisms in the soil. *Aust. J. Agric. Res.* 8:659–673.

Russell, E.W. 1971. Soil structure: Its maintenance and improvement. *J. Soil Sci.* 22:137–151.

Russell, E.W. 1973. *Soil Conditions and Plant Growth*, 10th Ed. Longman, London.

Russell, E.W. 1975. Reduced cultivations and direct drilling: The present position and the research needs of these techniques. *Outlook Agric.* 8:257–259.

Russell, M.B., A. Klute, and W.C. Jacob. 1952. Further studies on the effect of long-time organic matter additions on the physical properties of Sassafras silt loam. *Soil Sci. Soc. Am. Proc.* 16:156–159.

Russell, R.S. 1977. *Plant Root Systems—Their Function and Interaction with the Soil.* McGraw-Hill, London.

Russell, R.S., and M.J. Goss. 1974. Physical aspects of soil fertility—the response of roots to mechanical impedance. *Neth. J. Agric. Sci.* 22:305–318.

Sartori, G., G.A. Ferrari, and M. Pagliai. 1985. Changes in soil porosity and surface shrinkage in a remolded, saline clay soil treated with compost. *Soil Sci.* 139:523–530.

Satchell, J.E. 1967. Lumbricidae. In: A. Burgess and F. Raw (eds.) *Soil Biology.* Academic Press, London, pp. 259–322.

Scotter, D.R., and P. Kanchanasut. 1981. Anion movement in a soil under pasture. *Aust. J. Soil Res.* 19:299–307.

Sedgley, R.H., and K.P. Barley. 1958. Effects of root growth and decay on the capillary conductivity of a sandy loam at low soil-moisture tension. *Soil Sci.* 86:175–179.

Sequi, P. 1978. Soil structure—an outlook. *Agrochemica* 22:403–425.

Shields, J.A., and E.A. Paul. 1973. Decomposition of ^{14}C-labelled plant material under field conditions. *Can. J. Soil Sci.* 53:297–306.

Sparling, G.P., M.V. Cheshire, and C.M. Mundie. 1982. Effect of barley plants on the decomposition of ^{14}C-labelled soil organic matter. *J. Soil Sci.* 33:89–100.

Springett, J.A. 1983. Effect of five species of earthworms on some soil properties. *J. Appl. Ecol.* 20:865–872.

Springett, J.A. 1985. Effect of introducing *Allolobophora longa* Ude on root distribution and some soil properties in New Zealand pasture. In: A.H. Fitter (ed.) *Ecological Interactions in Soil: Plants, Microbes and Animals.* Blackwell's Scientific, London, pp. 399–405.

Stirk, G.B. 1954. Some aspects of soil shrinkage and the effect of cracking upon water entry into the soil. *Aust. J. Agric. Res.* 5:277–290.

Syers, J.K., and J.A. Springett. 1984. Earthworms and soil fertility. *Plant Soil* 76:93–104.

Tippkötter, R. 1983. Morphology, spatial arrangement and origin of macropores in some hapludalfs, West Germany. *Geoderma* 29:355–371.

Tisdall, J.M. 1985. Earthworm activity in irrigated red-brown earths used for annual crops in Victoria. *Aust. J. Soil Res.* 23:291–299.

Tisdall, J.M., and J.M. Oades. 1979. Stabilisation of soil aggregates by the root systems of ryegrass. *Aust. J. Soil Res.* 17:429–441.

Tisdall, J.M., B. Cockroft, and N.C. Uren. 1978. The stability of soil aggregates as affected by organic materials, microbial activity and physical disruption. *Aust. J. Soil Res.* 16:9–17.

Towner, G.D. 1986. Anisotropic shrinkage of clay cores, and the interpretation of field observations of vertical soil movement. *J. Soil Sci.* 37:363–371.

Unger, P.W., and D.M. Van Doren (eds.) 1982. *Predicting Tillage Effects on Soil Physical Properties and Processes*, Spec. Pub. 44. American Society for Agronomy, Madison, WI.

Utomo, W.H., and A.R. Dexter. 1981. Tilth mellowing. *J. Soil Sci.* 32:187–201.

Whiteley, G.M., and A.R. Dexter, 1983. Behaviour of roots in cracks between soil peds. *Plant Soil* 74:153–162.

Wiersum, L.K. 1957. The relationship of the size and structural rigidity or pores to their penetration of roots. *Plant Soil* 9:75–85.

Wiersum, L.K. 1961. Utilisation of soil by the plant root system. *Plant Soil* 15:189–192.

Distribution, Properties, and Management
of
Vertisols of India

A.S.P. Murthy[*]

[*]Current mailing address: University of Agricultural Sciences, Regional Research Station, V.C. Farm, Mandya—571 405, India.

© 1988 by Springer-Verlag New York Inc.
Advances in Soil Science, Volume 8

I. Introduction

Vertisols comprise a group of clay-textured soils that occur extensively in the tropics and temperate zones. They are synonymous with black cotton soils, black earths, dark clays, grumusols, and regurs of earlier classification systems (Dudal, 1965).

Vertisols are typically dark colored and are characterized by a high percentage of clay dominated by a smectite group of minerals. Dominance of smectite group of minerals leads to expansion and shrinkage on wetting and drying. Low infiltration rate, high plasticity and stickiness, low organic matter, high cation exchange capacity, calcareous nature, and alkaline reaction are some of the properties associated with these soils.

Vertisols are extensively cultivated, and their productivity makes a significant contribution to the national economy of India. However, there remain large areas of vertisols that are underutilized partly owing to inherent management problems and partly because of deficiencies in supplies of irrigation water and failure to recognize nutrient limitations.

It is the purpose of this review to present a summary of available knowledge on the distribution of vertisols in India, their problems and potentialities for agricultural production and on management technology to alleviate soil-crop constraints for increased crop production.

Most of the examples are drawn from the coordinated projects on soil and crops, International Crops Research Institute for the Semi-Arid Tropics (ICRISAT), and the Advance Centre for Black Soil Research (ACBSR), India, reflecting the author's experience without drawing extensively on related work performed in other parts of the world.

This review is mainly devoted to vertisols of India. The discussion in this review will be of use where vertisols exist and are used for many of the same crops and require the same or very similar management for successful crop production. Soil taxonomy terminology (Soil Survey Staff, 1975) for vertisols will be used in this review.

It is hoped that with the adoption of new technology vertisols can produce many times more food than they produce today.

A. Distribution

It is estimated that at least 275 million hectares of vertisols and associated vertic soils occur throughout the world. Extensive areas of these soils are located in Australia, India, Sudan, Chad, Ghana, Cuba, Puerto Rico, the United States, and many other countries including Venezuela, Ethiopia, Egypt, Syria, and Turkey (Figure 1) (Dudal, 1965).

Vertisols and associated vertic soils in peninsular India occur mostly extending from 8° 45' to 26° 0' N latitude and 66° 0' to 83° 45' E longitude, covering an area of 73 million hectares. The area occupied by vertisols constitutes nearly 22.2% of the total geographical area of India. Distribution of vertisols in different states of India is given in Table 1 (Murthy, 1981).

B. Major Constraints and Potentialities

The main soil-related problems in vertisols are low fertility, limited soil moisture availability, and poor drainage. Even with these constraints, vertisols have great potential for crop production because of high clay content resulting in a substantial storage of water, which allows crops to grow and survive for longer periods. Sinha and Swaminathan (1979), from a theoretical analysis of climate, soil, and water resources of India, calculated the maximum production potential in terms of grain equivalents, and it is of

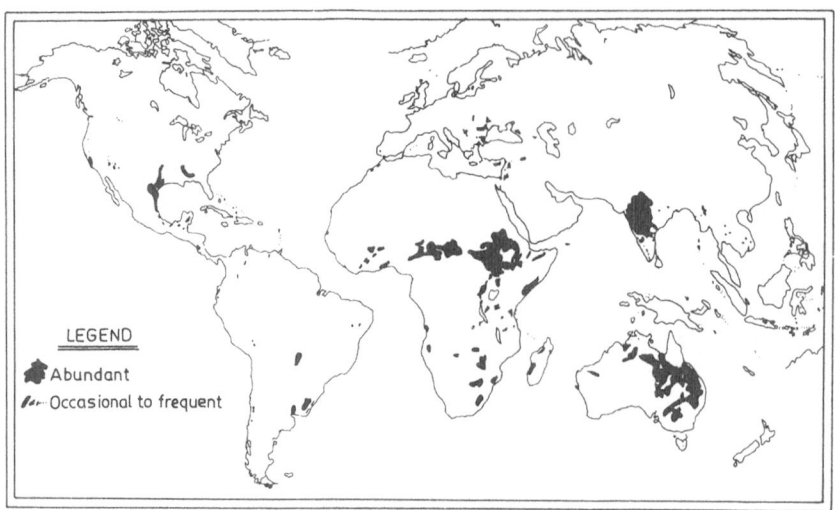

Figure 1. Worldwide distribution of vertisols (FAO-UNESCO Soil Map of the World 1971–1979).

Table 1. Distribution of vertisols and associated soils in India

State	Total area under vertisols and associated soils (million ha)	Area under vertisols and associated soils expressed as	
		Gross vertisols area in India (%)	Total geographical area in India (%)
Maharashtra	29.9	35.5	7.9
Madhya Pradesh	16.7	23.0	5.1
Gujarat	8.2	11.9	2.6
Andhra Pradesh	7.2	10.0	2.2
Karnataka	6.9	9.4	2.1
Tamil Nadu	3.2	4.2	1.0
Rajasthan	2.3	3.0	0.7
Orissa	1.3	2.0	0.4
Bihar	0.7	1.0	0.2
Uttar Pradesh	Negligible	Negligible	Negligible

Source: Murthy (1981).

the order of 1235×10^6 t yr^{-1} in vertisols. The results shown in Figures 2 and 3 support the theoretical calculation and indicate the scope for increasing crop yields in farmer's fields (AICRPDA, 1982). This apparently shows that availability of moisture and/or supplemental irrigation is the determining factor to increase the yield when fertilizer (30 kg N/ha^{-1}) is applied (Figure 4).

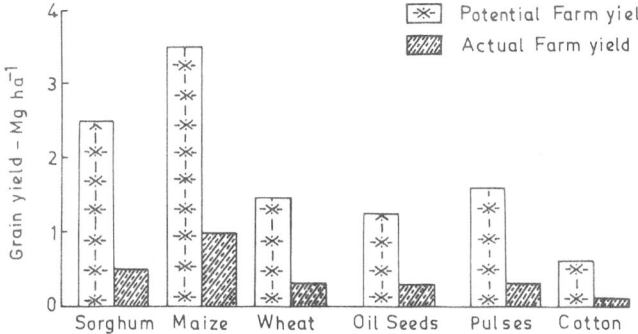

Figure 2. Yield gaps in dry land agriculture on vertisols (AICRPDA, 1977).

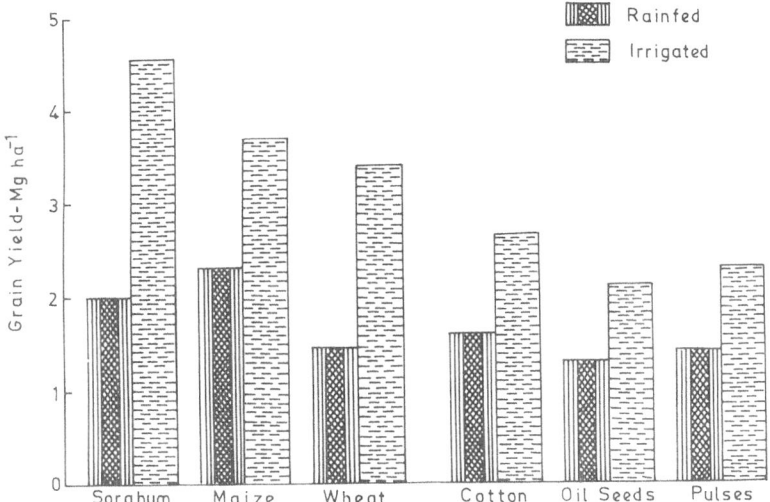

Figure 3. Yield potentials of various crops under rain-fed and irrigated conditions (AICRPWM, 1981).

II. Physical Properties

A. Texture

The textures of vertisols are generally clay loam, silty clay loam, silty clay, and clay. The clay content ranges from 40% to 80%. The clay content remains uniformly high (>35%) throughout the profile to a depth of at least 500 mm or more (Raychaudhuri *et al.*, 1963). In some vertisols where the topsoil is probably eroded, the clay content may be less than 40%, leading to loam or silty loam texture.

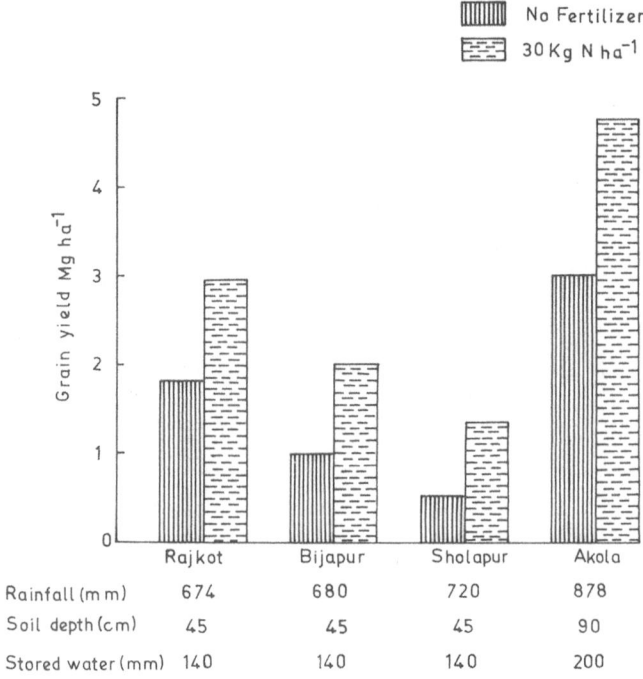

Figure 4. Grain yield of sorghum as affected by stored water and nitrogen (AICRPDA, 1982).

B. Soil Depth

Raychaudhuri *et al.* (1963) classified the vertisols of India into the following categories on the basis of soil depth. The depth varies from shallow to very deep, although the texture is generally fine in all cases.

1. Deep black soils
2. Medium and light black soils
3. Shallow black soils

The medium, 450- to 900-mm depths, and deep, greater than 900-mm black soil profiles generally occur on the flatlands and valleys with very gentle slopes. The soils on the highest position of the catena (hillocks) are characterized by a shallow depth, less than 450 mm. About 50% of the vertisols are shallow or medium in depth.

C. Soil Moisture

Available soil water storage capacity and the likelihood of moisture stress are dominant factors affecting the crop growth in vertisols, since crop

Table 2. Length of growing season and cropping intensity in vertisols at a rainfall probability of 75%

Soil depth (cm)	Available water-storage capacity (mm)	Rainfall (mm)	Effective cropping season (weeks)	Soil moisture availability period (weeks)	Cropping intensity
Up to 45	140	650	< 20	21	Sole cropping
45–90	200	700	20–30	25	Inter-cropping
90	300	800	> 30	31	Sequence cropping

Source: Virmani *et al.* (1978).

yields are largely proportional to moisture availability. Available water storage is largely determined by soil depth and the intensity and duration of rainfall. The vertisols have a high content of expansible layer silicates and therefore a high water storage capacity, ranging from 140 to 300 mm. The stored moisture availability period ranges, from 21 to 31 weeks (Table 2). Besides soil texture, type of clay, organic matter content, and soil depth, available soil water storage depends on the type, surface conditions, and moisture status of the soil.

Soil moisture recharge and utilization are very much controlled by potential evaporative demands and rainfall. Krantz *et al.* (1977) have developed a calendar for raising crops on the basis of weather, soil profile moisture characteristics, and evaporation data. Considering this, it appears that cropping season and cropping intensity depend on the moisture-holding capacity of the soil (Table 2).

Even though the soils are said to be highly retentive of moisture, the moisture held at permanent wilting point (PWP) estimated by 15 bar suction is quite high, ranging from 0.22 to 0.28 kg/kg^{-1}. This means that substantial rainfall is required to bring the water content of a dry soil above the wilting point. Since the precipitation in the early part of the monsoon is quite inadequate, the medium-deep soils (beyond 450 mm depth) usually do not have adequate moisture for sowing. It is only in the month of September, after receipt of about 200 mm of rains, that the medium and deep soils are adequately moistened for cropping. Hence, cropping after the rainy season is predominant in vertisols.

The soil depth often determines the crop growth. A medium to long duration crop can be grown on a deeper soil, as opposed to a short duration crop on a low-water-holding soil with a shallower depth. Since soil types and rainfall patterns show considerable variations over short distances, an

Table 3. Infiltration rates of vertisols

Location	Infiltration rates, mm/h^{-1}, after:				
	0.083 h	0.5 h	1 h	2 h	2.4 h
ICRISAT,[a] Hyderabad	—	76	34	4	0.2
Rahuri,[b] Maharashtra	14	—	11	9	—

[a] Krantz et al. (1978).
[b] Magar, (1982).

analysis of these factors can assist considerably in deriving estimates of the crop-growing periods and suitable crops and in delineating the crop-growing environment at specified benchmark locations. This will also assist the plant breeders in the identification of zones for which suitable crop varieties need to be bred and in determining the required characteristics. With substantial refinement, the methodology can be applied to many regions. It also allows for the identification of ecological isoclimes, making more feasible the transfer of appropriate farming from one region to another.

D. Infiltration Rates

Infiltration rates for vertisols are low to moderate. The results shown in Table 3 for two vertisols indicate high initial infiltration rates which drop sharply after the first hour and remain low when the soil gets saturated (Krantz et al., 1978 and Magar, 1982). The formation of a thin crust or a dispersed layer at the surface due to the impact of falling raindrops seems to be one of the factors in lowering infiltration rates after the first hour.

During the rainy season, infiltration rate and saturated hydraulic conductivity are low, resulting in impeded drainage which affects adversely the air-water relations necessary for proper land preparation and satisfactory crop establishment at the right time.

E. Soil Erosion

Vertisols in India are located on either relatively flat or sloping lands. Soil and water erosion are serious problems in these soils because of high rainfall intensity, lack of vegetative cover when the rainy season commences, and poor infiltration characteristics, resulting in a soil loss averaging around 6 $t/ha^{-1}/yr^{-1}$ (Kanwar et al., 1982).

Table 4. Chemical properties of vertisols

Properties	Location				
	India[a]	USA[b]	Africa[c]	Israel	Australia[c]
PH	8.4	7.7	8.1	7.5	7.4
Organic carbon g/kg^{-1}	7.0	19.0	16.0	—	28.0
Cation exchange capacity mmol/kg^{-1}	650	582	540	270	724
CaCO$_3$ g/kg^{-1}	151	330	150	370	—

[a] ACBSR (1981).
[b] G.W. Kunze (1985, personal communication).
[c] Dudal (1965).

III. Chemical Properties

Some important data on chemical properties of vertisols occurring in different parts of the world are presented in Table 4.

A. pH

The pH values of vertisols range between 7.4 and 8.1, indicating that the chemical nature of vertisols occurring in diverse environments are more or less similar. Generally, pH values are found to increase with depth, corresponding to an increase in CaCO$_3$ and salts of calcium, magnesium, and sodium. Vertisols that have been irrigated and/or those occurring in depressions are found to have pH values as high as 9.5 at the surface.

B. Organic Carbon

The organic carbon contents of vertisols are relatively low with a large variation when compared with figures from different countries. In vertisols of India, organic carbon content of the surface soil ranges from 7 to 10 g/kg^{-1}. According to Dudal (1965), the content of organic matter varies from 5 to 20 g/kg^{-1} in most of the African vertisols and from 20 to 40 g/kg^{-1} in some vertisols from the United States. Some Australian black earths have been reported to contain organic matter as high as 60 g/kg^{-1} at the surface.

Organic carbon content has been found to be more or less uniformly distributed up to a depth of 1 or 2 m in some vertisols of Karnataka. For example, organic carbon in the profile of a deep vertisol decreased from 7.5 to 6 g/kg^{-1} (ACBSR, 1980).

The organic matter buildup in vertisols appears to be governed by existing natural vegetation and cropping history and temperature (Dudal, 1965; Jenny and Raychaudhuri, 1960).

C. Cation Exchange Capacity

Cation exchange capacity (CEC) values for the surface horizons of vertisols range from less than 270 to more than 720 mmol/kg^{-1} soil. CEC does not appear to change much with depth, suggesting the major role played by inorganic soil component rather than the organic matter (Dudal, 1965; ACBSR, 1982). Vertisols are having higher CEC because of the dominance of expansible layer silicates.

D. Calcium Carbonate

Calcium carbonate occurs in vertisols either as concretions or in a finely powdered form. The content of $CaCO_3$ may vary from nil to 330 g/kg^{-1}. In vertisols of India, it ranges from 5 to 200 g/kg^{-1}; in Australian vertisols, from 50 to 150 g/kg^{-1}; and in the vertisols of Java, from 50 to 600 g/kg^{-1} (Dudal, 1965). In calcareous soils, solid-phase $CaCO_3$ is believed to govern P reactions. A recent study (ACBSR, 1985) reveals that the P sorption characteristics of vertisols is not well related to total $CaCO_3$, since the reactivity of $CaCO_3$ is dependent on specific surface, which is related to carbonate particle size distribution, rather than the chemically determined $CaCO_3$.

IV. Mineralogical Properties

A. Clay Mineralogy

It is generally believed that the clay mineralogy of vertisols is dominated by smectite. But the results of X-ray diffraction analysis work of the Advance Centre for Black Soil Research (ACBSR, 1981) on a large number of vertisol clays of Karnataka, India, revealed the presence of chloritized iron-rich smectite (Figures 5, 6), with practically no kaolinite and mica. This finding is in contrast to that of Krishna Murti and Satyanarayana (1969, 1970); Singh and Krishna Murti (1974); and Chatterjee and Rathore (1976) who reported an abundance of smectite with mica and kaolinite in vertisols. Further, Singh and Krishna Murti (1975) reported the presence of beidellite-nontronite minerals in vertisols of Mandasur, Madhya Pradesh and Poona, Maharashtra, India. The chlorite component in smectite occurring in vertisols of Karnataka as reported in this investigation appears to be

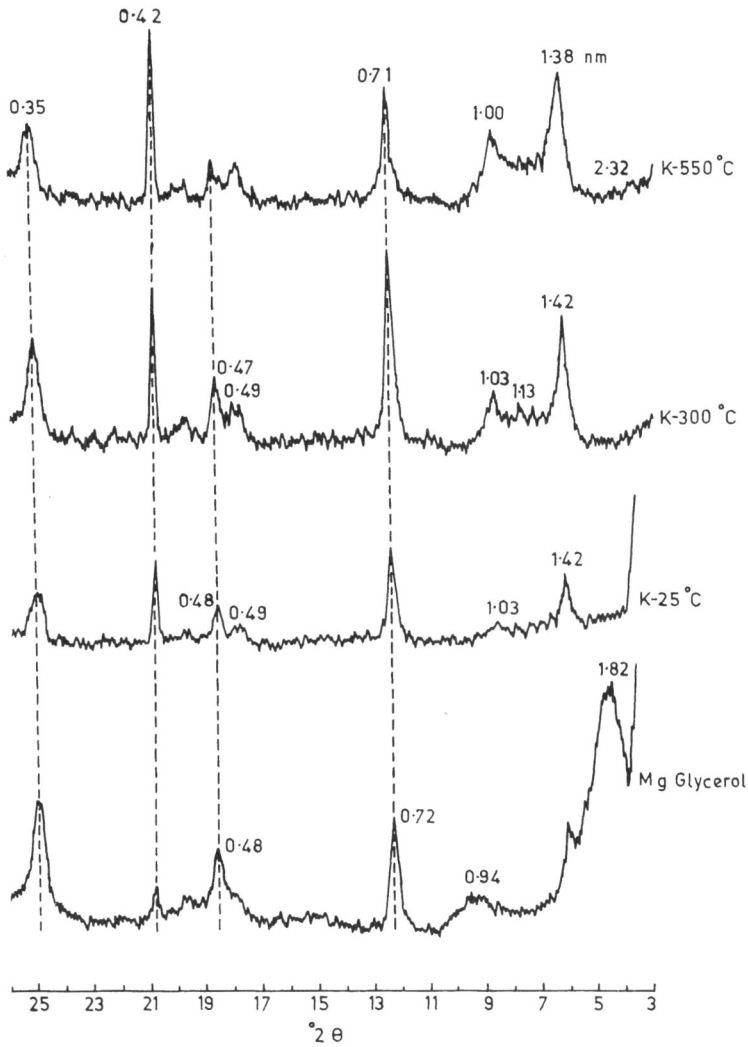

Figure 5. X-ray diffractograms of vertisol (Typic Chromusterts) clay fraction (2–0.2 μm).

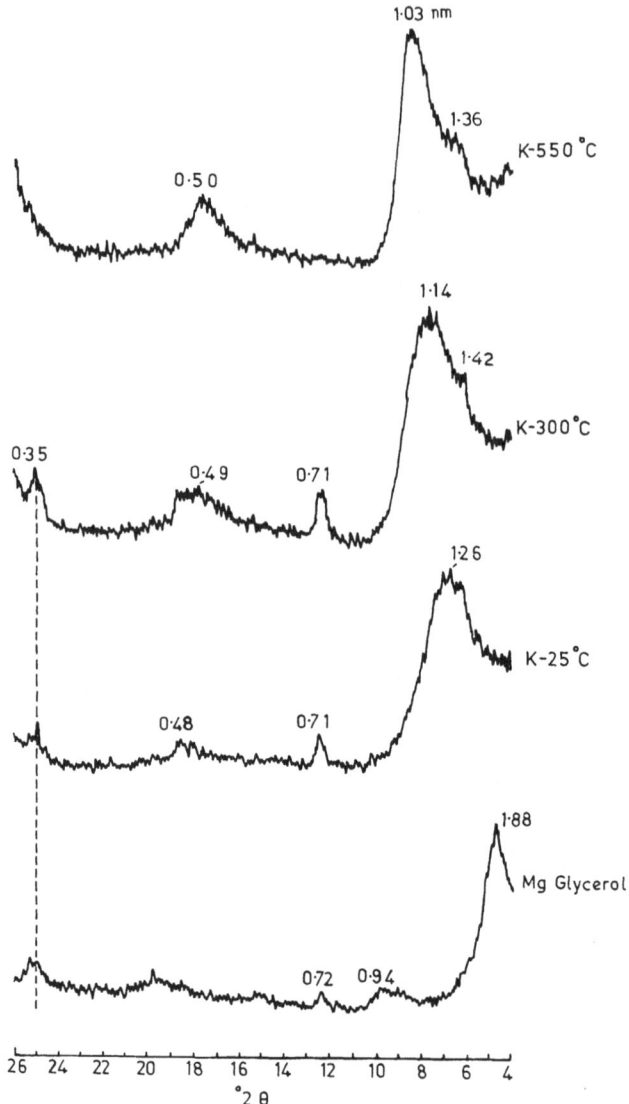

Figure 6. X-ray diffractograms of vertisol (Typic Chromusterts) clay fraction (less than 0.2 μm).

a product of precipitation of hydroxy aluminium in smectite interlayers. The response to heat treatment of K-saturated samples as shown in Figures 5 and 6 clearly indicated the presence of hydroxy-interlayered forms, since hydroxy-interlayered minerals do not collapse exactly 1.0 nm upon K saturation and subsequent heat treatment. These hydroxy-interlayered clays are known to influence a wide range of physical and chemical reactions in soils, including shrink and swell, degree of dispersion, and adsorption of phosphorus and micronutrients.

Elemental analysis of the clay fraction less than 0.2 μm indicates that clays separated from vertisols of Karnataka have charge located mainly in the tetrahedral layer. The structural formula of the typical soil smectite is

$$(Si_{7.3}Al_{0.7})^{IV}(Al_{2.3}Fe_{0.8}Mg_{0.9})^{VI}O_{20}(OH)_4$$

The clay mineralogy of the vertisols of Karnataka suggests that they possess high layer charge.

B. Sand and Silt Mineralogy

Most of the soil mineralogy research studies in India have been directed toward the identification of clay minerals (Ghosh and Kapoor, 1982), but few data are available on the mineral suites in the sand and silt fractions of vertisols. X-ray diffraction analysis of the fine sand fraction following heavy liquid separation at a specific gravity of 2.96 using 1,1,2,2-tetrabromoethane into light and heavy mineral fractions revealed the presence of alkali feldspars including orthoclase, microcline, and albite in vertisols of Karnataka. X-ray diffraction analysis of the silt fraction showed the presence of 1.4- and 1.0-nm layer silicates in addition to nonclay minerals such as alkali feldspars (ACBSR, 1982). From the foregoing discussion, it is clear that the smectite in vertisols of Karnataka might have formed as a product of weathering of feldspars.

V. Technology to Alleviate Adverse Soil Physical Properties

A. Broad Bed and Furrow System

Recent research at ICRISAT (Kanwar et al., 1982; El-Swaify et al., 1985) shows that the broad bed and furrow system (BBF), use of improved implements, high-yielding varieties, balanced use of fertilizers, and improved crop management practices have resulted in significant reduction in soil and water loss (Table 5).

Hydrologic studies of the traditional rainy season fallowing have shown that of the total rainfall potentially available, about 25% is lost through runoff, 25% is lost through evaporation from the bare fallow soil, 10% is lost through deep percolation, and only 40% is actually utilized for evapotranspiration by a post-rainy season crop (Kanwar et al., 1982).

Table 5. Estimated water balance components of a deep vertisol in the rainy season fallow and maize-chickpea sequential cropping system at ICRSAT

Water balance component as % total rainfall	Rainy-season fallow 1973–78 (mm)	Improved technology[a] 1977–79 (mm)
Runoff	25.3	12
Deep percolation	9.2	10
Evaporation	24.9	6[b]
Evapotranspiration	40.6	72[c]
Rainfall	785	872
Soil loss (Mg/ha^{-1})	6	1.5

Source: Kanwar *et al.* (1982).

[a] Broad bed and furrow combined with sequential cropping systems, high-yielding varieties, and balanced use of fertilizers.

[b] During dry season.

[c] During crop season.

A comparison of the water balance components in the traditional rainy season fallow with that of the improved technology clearly shows that the surface runoff is reduced by more than 50% and evaporation loss is reduced by over 75%, while evapotranspiration increased from 41% to 73%, indicating greater availability of water for production purposes. The improved management of deep vertisols greatly reduced the erosion losses to less than one-fourth of the fallow treatment (Table 5).

BBFs on medium and shallow vertisols do not significantly affect runoff or erosion, nor do they substantially increase yields. This seems to be due to the absence of serious surface and subsurface drainage problems on the medium and shallow vertisols (ICRISAT, 1981). More research is necessary to understand the differing effects of BBF on medium to shallow, compared to medium-deep and deep, vertisols.

B. Land Configuration

Gupta *et al.* (1978) studied the effect of land configuration on a black clayey soil of Madhya Pradesh with soybean (*Glycine max* (L.) Merr) as a test crop. The land treatments were 3, 6, and 9 m wide, 200 mm high, raised beds. Data on soybean yields recorded over three rainy seasons (1974–76) showed a 14% increase in yield on 3-m beds over that of 9-m beds. These land treatments provided adequate surface drainage for the test crop and enhanced the permeability of water under high-rainfall conditions. Grain yields of gram (*Cicer arietinum* L.) followed by soybean and wheat (*Triticum durum* Dest L.) followed by rice (*Oryza sativa* L.) were 2.2 and 1.7 Mg/ha^{-1}, respectively. This demonstrates the need for adoption of land configuration systems leading to speedy disposal of excess water in stabilizing agricultural production in rain-fed areas (Table 6).

Table 6. Grain yield of soybean (Mg/ha^{-1}) as affected by land configurations

Width of upland plots	Year		
	1974[a]	1975[b]	1976[b]
3 m	2.2	3.9	3.3
6 m	1.8	2.7	2.9
9 m	1.8	2.7	2.9
CD (1%)	0.598	0.205	0.160
CD (5%)	0.322	0.137	0.116

Source: Gupta *et al.* (1978).

[a] Average of five replications.

[b] Average of eight replications.

Table 7. Influence of various mechanical structures on grain yields in vertisols

Crop	Average yields (Mg/ha^{-1})			
	Unbunded	Contour bunds	Graded bunds	Broad-based terraces
Postrainy sorghum	0.24	0.18	0.28	0.36
Cotton	0.13	0.48	1.65	1.10
Safflower	0.19	0.12	0.21	0.58

Source: Chittaranjan *et al.* (1980).

C. Mechanical Structures

Research studies conducted at Bellary by Chittaranjan *et al.* (1980) showed that contour bunds are highly unsuited to vertisols because of water stagnation and breaches of bunds. Among the various structures evaluated, the drainage-type terraces (namely, graded bunds) proved most efficient. Bunds of 0.8 m^2 cross section at vertical intervals of 0.7 m with a channel (grade of 0.1–0.25%) on the upstream side connected to a grassed waterway have been found useful. These structures not only brought down soil losses from 12 Mg/ha^{-1}/yr^{-1} to around 1 Mg/ha^{-1}/yr^{-1} but also increased the crop yields considerably (Table 7).

D. Tillage

Tillage with the objective of preparing a good seedbed and moisture conservation often helps in the absorption of moisture to greater depths. Different tillage operations coupled with farmyard manure (FYM) were

Table 8. Effect of tillage operations on moisture storage in the soil profile (mm)

Date of observation	Rainfall (mm)	Blade harrow (10 cm)	Heavy-duty disk (15 cm)	Subsoiling (45 cm)	Basin scooping	Scooping	Plowing	
							Shallow (20 cm)	Deep (30 cm)
July 27, 1975	163	97	102	119	130	132	110	119
August 16, 1975	360	262	257	269	264	267	263	266
September 10, 1975	473	312	302	301	301	305	303	308

Source: Patil *et al.* (1981).

Table 9. Effect of vertical mulch on yields of sorghum (Mg/ha^{-1})

Interval of vertical mulch	Location		
	Bellary (1973–76)	Sholapur (1974–76)	Bijapur (1973–74)
No mulch	0.836	0.840	1.661
2 m	1.172	1.237	—
4 m	1.281	1.266	2.050[a]
8 m	1.186	1.123	1.823[a]

Source: Randhawa and Rama Mohan Rao (1981).

[a] 5 and 10 m spacing.

tested by Patil *et al.* (1981) at the Dry Farming Centre, Sholapur, to study the moisture conservation.

The data in Table 8 indicate that as the rainfall increases, the moisture storage in the soil profile becomes more or less uniform, irrespective of the tillage operations. In the low-rainfall period, deep tillage and scooping helped in better moisture storage. Deep tillage, even though effective in the beginning, could not increase soil water storage at the end because of limitations of soil profile depth.

An experiment conducted at Hagari, Karnataka, showed that bunding and scooping not only helped in greater retention of moisture but also facilitated in better *in situ* harvest of rain water as compared with the control. The retention values of moisture due to bunding and scooping were 59% and 81%, respectively, as compared with plots that were not bunded and scooped with basin lister (Gopalkrishna Rao *et al.*, 1975).

E. Vertical Mulching

In the Bellary region, where the rainfall is about 500 mm, the infiltration rate was low, leading to substantial runoff and erosion even when the exchangeable sodium percentage was only 8. In such situations, it was found that by keeping sorghum (*Sorghum bicolor* L.) stubbles as vertical mulch in trenches 40 cm deep, 150 cm wide across the slope at 2-, 4-, and 8-m intervals extending 10 cm above the ground level enhanced the available soil moisture by 40–50 mm. This led to better *in situ* water harvesting and higher grain yields (Table 9). Vertical mulching has improved water intake into heavy black soil, thereby leading to better availability of nutrients (Randhawa and Rama Mohan Rao, 1981).

F. Organic Residues and Infiltration Rates

Table 10 shows the results of Magar (1982) for vertisols of Rahuri, Maharashtra, where incorporation of organic residues was effective in increasing

Table 10. Effect of organic residues on the infiltration rate in vertisols[a]

| | Infiltration rate (mm/h^{-1}), after: | | |
Treatment	15 min	60 min	120 min
Control	14	11	9
Farmyard manure (20 Mg/ha^{-1})	40	35	32
Press mud (10 Mg/ha^{-1})	30	17	15
Wheat straw (5 Mg/ha^{-1})	27	18	16
Wheat straw (10 Mg/ha^{-1})	32	22	18

Source: Magar (1982).

[a] Infiltration rate was measured at an initial moisture content (0- to 15-cm layer) of 0.35 kg/kg^{-1}.

Table 11. Effect of advancing sowing dates on grain yields (Mg/ha^{-1}) at different regions in black soil areas

Region	Crop	Date of sowing (day/mo)	Yield (Mg/ha^{-1})
Bijapur	Jowar	08/09	3.12
		23/09[a]	1.91
	Safflower	08/09	2.89
		23/09[a]	1.31
Sholapur	Jowar	25/08	1.87
		02/10[a]	1.40
Bellary	Jowar	19/09	4.95
		12/10[a]	2.00
Indore	Sunflower	22/08	1.89
		22/09[a]	0.82

Source: S.L. Chowdhury (1978).

[a] Conventional sowing.

the basic infiltration rate. Incorporation of press mud (10 Mg/ha^{-1}) and wheat straw at the rate of 5 Mg/ha^{-1} resulted in a twofold increase in the infiltration rate over the control.

G. Sowing Dates

Table 11 shows the results of AICRP on dry land agriculture on the effects of sowing dates on the yield of crops. Early sowing of crops helps in obtaining good and vigorous seedlings. Advancing the sowing time of rabi crops has resulted in enormous yield advantage over conventional time of sowing. The yield advantage due to advance in sowing time seems to be primarily due to optimized use of stored moisture during grain filling stages (Chowdhuary, 1978).

H. Fertilizers on Stored Soil Moisture/Irrigation Use Efficiency

The role of fertilizer in increasing grain yield and in affecting the efficiency of rainwater use (WUE) and where available, supplemental irrigation water use (WUE-I), under rain-fed farming, though important, has not received adequate attention. The data in Table 12 illustrate the use of nitrogen on WUE in the stored moisture environment, and the efficiency of irrigation.

In a field experiment on deep vertisols, Kanwar et al. (1984) reported that application of 80 kg N/ha^{-1} increased the consumptive water use by 31% over no fertilizer. Irrigation increased the grain yield and also significantly increased the consumptive use of water. The WUE-I was almost twofold in 1980 and fivefold in 1981 with 80 kg/N/ha^{-1} as compared to no nitrogen application. The authors explained the differences in grain yield on the basis of NO_3-N in the top 120 cm soil at the time of planting of the sorghum crop. During 1980, NO_3-N was 40 kg/ha^{-1}, and in 1981 it was 13 kg/ha^{-1} (Table 12).

In a similar study, Hanwante et al. (1981) studied the effect of N and P rates on yield and WUE in a medium black soil at Akola with safflower (*Carthamus tinctorius* L.) as a test crop. The data in Table 13 show that application of N increased the WUE significantly. The increase in WUE was 37% and 47% where 25 kg and 50 kg/N/ha^{-1} were applied, respectively.

The data obtained by Kanwar et al. (1984) and Hanwante et al. (1981) clearly suggest that fertilizer application is essential for making best use of stored moisture and/or irrigation water. Fertilizer application has helped in transporting more nutrients into the plant, which leads to higher yield.

Table 12. Effect of N on water use efficiency in a stored moisture environment and efficiency of irrigation

N applied (kg/ha^{-1})	No. of irrigations	Sorghum grain yield (Mg/ha^{-1}) 1980	1981	Consumptive water use (mm/ha^{-1}) 1980	1981	WUE-C (kg/mm^{-1}/ha^{-1}) 1980	1981	WUE-I (kg/mm^{-1}/ha^{-1}) 1980	1981
0	0	3.458	1.083	114.5	147.5	3.02	0.73	—	—
	2	4.035	1.541	268.5	220.5	1.50	0.70	4.4	3.5
80	0	4.053	2.507	150.5	201.5	2.81	1.25	—	—
	2	5.319	4.764	335.5	290.5	1.60	1.65	9.7	17.4

Source: Sardar Singh and Seetharama, unpublished (cited by Kanwar et al., 1984).

Table 13. Effects of N and P rates on yield and WUE of safflower crop

Treatments (kg/ha^{-1})	Grain yield (Mg/ha^{-1})	Total water used (mm)	WUE (kg/mm^{-1}/ha^{-1})	FUE[a]
Nitrogen				
0	0.923	251.04	3.67	—
25	1.273	251.04	5.05	14.0
50	1.391	258.29	5.38	9.3
CD (5%)	0.244		0.29	
Phosphorus				
0	1.123	251.24	4.43	—
10.9	1.236	265.80	4.65	4.5
21.8	1.251	249.80	5.01	2.5

Source: Hanwante *et al.* (1981).

[a] kg grain kg^{-1} fertilizer nutrient.

From the foregoing, it is evident that the high water storage capacity of vertisols for producing several times the crop yields currently achieved can be realized by adopting the following strategy: Climate-water balance studies to help in (1) the selection of plant type, cropping system, and synchronization of crop growth period with water cycle; (2) adequate water conservation within the soil profile or efficient utilization of rainfall and stored water by adopting proper land configurations, tillage, mulches, and use of fertilizers.

VI. Nitrogen

Since nitrogen is the fertilizer nutrient most often needed by plants and is also the most vulnerable to losses due to volatilization and leaching, major emphasis is focused on nitrogen forms and various aspects of nitrogen fertility management.

A. Nitrogen Forms in Vertisols

The total N content in vertisols of India is generally about 0.8 g/kg^{-1}. The distribution of total N content in the profile closely follows that of organic carbon.

Organic carbon and alkaline permanganate oxidizable N are the most used methods for evaluating N fertility status of Indian soils (Ghosh and Hasan, 1980). Recent work by Sahrawat and Burford (1982) indicates that the alkaline permanganate digestion method, which has been widely used, is not capable of measuring one of the useful forms of N in vertisols for plant growth—i.e., nitrate. Nitrate nitrogen is found in appreciable amounts (3.3–27.7 mg kg^{-1}) in vertisols that experience pronounced wet and dry seasons (Shukla and Singh, 1968).

Recent work by Rego et al. (1982) shows that nitrates could accumulate to the extent of 30–40 mg kg^{-1} in the upper 1 m of the vertisol profile which decreased to 6–9 mg kg^{-1} under monsoon cropping. This is of practical significance in seasonally dry soils like vertisols, as there is a flush of nitrate formation at the onset of monsoon following a hot and dry spell. The longer the dry period, the higher the flush of mineral and nitrate nitrogen. Perhaps planting of crops to tap the flush of mineral nitrogen is an important consideration in the semiarid tropical regions (ICRISAT, 1978). These results apparently suggest that NO$_3$-nitrogen is an index of N availability in vertisols and requires further study.

B. Factors Affecting N Content in Vertisols

1. Crops and Cropping Systems

The study by Jenny and Raychaudhuri (1960) clearly indicated that the total N content in soils was dependent on temperature, rainfall, and altitude. Mutatkar and Raychaudhari (1959) observed that in black soils total N increased from 0.043% to 0.072% as the rainfall increased from 240 to 1000 mm.

In a long-term study initiated during 1930 at the Agricultural College, Poona, cereal-legume rotations and mixtures have brought about important changes in available N and organic carbon status in vertisols (Patil et al., 1982). At the end of 42 years of cropping, it was found that the total nitrogen content in the soil increased by 500 kg/ha^{-1} owing to inclusion of legumes like soybean and groundnut (Arachis hypogaea L.) in the rotation with pearl millet (Pennisetum americanum (L.) Leeke). This works out to about a 50% increase in available nitrogen compared to that in cereal after cereal rotation (Table 14).

Recent work by ICRISAT (1980) has shown that the cultivation systems affect the grain yield and uptake of N by sorghum (Sorghum bicolor (L.)

Table 14. Available N and organic carbon contents in vertisols (0–150 mm depth) as affected by cropping systems

Treatment	Available N (kg/ha^{-1})	Organic carbon (g/kg^{-1})
Pearl millet after pearl millet	468	6.4
Pearl millet after mung	624	6.7
Pearl millet after soybean	578	7.3
Pearl millet plus pigeon pea continuous	484	6.7
Pearl millet after groundnut	544	6.4

Source: Patil et al. (1982).

Table 15. Effect of different cultivation systems on yield and uptake of N by sorghum at ICRISAT, 1980

	Cultivation method[a]				CV (%)
	Zero	Shallow	Deep	BBF	
Grain yield (Mg/ha^{-1})	1.82	2.00	2.09	2.06	3.8
Grain + straw (Mg/ha^{-1})	7.70	8.58	9.00	8.93	2.9
Nitrogen uptake in grain	18.5	21.2	23.4	22.5	6.7
Nitrogen uptake in grain + straw	59.0	67.9	77.6	76.6	6.8

Source: ICRISAT (1981).

[a] Zero, shallow, and deep are the three tillage treatments of 0-, 50- and 100-mm depths.

Moench) (Table 15). The data in Table 15 show that cultivation systems involving deeper and more thorough disturbance of the soil would promote mineralization of soil organic matter and a greater supply of mineral nitrogen for crop uptake. Preliminarily tillage treatments consisting of 0-, 5-, and 10-cm depths and the BBF system indicate that an increase in intensity of cultivation caused a small increase in grain yield and increased the nitrogen uptake by the crop.

2. Fertilizers and Manures

Addition of nitrogen to soils originates from rain and dust, symbiotic and nonsymbiotic fixation, and animal and human wastes. Application of fertilizers in addition to organic manures is often essential for increased crop production. Information on the distribution of N in various forms as affected by continuous application of manures and fertilizers is scanty.

The study by Puranik *et al.* (1978) on the continuous use of manures and fertilizers for more than 8 years shows that FYM alone and/or with green manure has a pronounced influence in increasing the total and organic nitrogen contents in vertisols by 56% over control (Table 16). The increase in the nitrogen fractions and total nitrogen was insignificant. It is because mineral fertilizers obviously caused greater decomposition and mineralization. It is also apparent from the data that 87% of the total N is in the organic fraction (Table 16). These results and those of Subbaiah and Sachdev (1981) using ^{15}N in a maize (*Zea mays* L.)-based multiple cropping system showed that crops utilized only 38% of applied N and nearly 62% of N immobilized in the profile.

Moraghan *et al.* (1984) found in field studies using ^{15}N-labeled fertilizer that residual soil N derived from fertilizer was of little value for subsequent

Table 16. Nitrogen fractions as affected by fertilizers and manures (g/kg^{-1} soil)

	Inorganic	Organic	Total
Control	0.094	0.615	0.70
FYM	0.136	1.018	1.12
NPK	0.092	0.644	0.70
FYM + green manures	0.178	0.903	1.05

Source: Puranik *et al.* (1978).

crops during the postrainy and rainy seasons. The authors believe that residual N is mainly organic N, a portion of which is probably associated with sorghum roots.

3. Losses of Nitrogen

The main avenues of N loss that are of practical significance are (1) volatilization losses of ammonia, (2) leaching as nitrate, and (3) denitrification.

Ammonia (NH_3) volatilization has received considerable attention in recent years with the rapid increase in the availability of urea in the fertilizer market. Urea has potential for NH_3 loss in both acid and calcareous soils; hence, discussion is restricted to ammonia volatilization in this review. The literature on ammonia volatilization has been reviewed by Tandon (1974), Sahrawat (1979), and Fenn and Hossner (1985). Among the factors that affect loss of ammonia through volatilization are pH, $CaCO_3$, temperature, cation exchange capacity, texture of the soil, fertilizer nitrogen source and its rate and method of application, and water regime of the soil.

In a laboratory experiment, More *et al.* (1977) studied the volatilization losses of ammonia from different N carriers such as urea, ammonium sulfate (AS), Uramphos (28N:12.2P), and Suphala (15N:6.5P:12.4K) in black soils of Marathawada. Volatilization losses of ammonia were found to increase with increase in pH of the soil. Ammonia volatilization was maximum during the 5-to-12-day period after the application of fertilizer (Figure 7). The ammonia loss was 12.5% from urea and lowest from Suphala. Ammonia volatilization losses decreased in the following order: urea > ammonium sulfate > Uramphos > Suphala. Studies by More *et al.* (1977) show that addition of well-decomposed FYM reduced the volatilization loss of ammonia from all the sources of N (Figure 8). Research carried out by Fenn and Kissel (1973) with Houston black clay found a maximum NH_3 loss of 55–65% with $(NH_4)_2SO_4$, $(NH_4)_2HPO_4$, and NH_4F at 22°C. Fenn and Kissel (1973) and Fenn and Miyamoto (1981) have also shown significantly less NH_3 loss from urea than from ammonium sulfate banded in soil at all depths. Ammonium sulfate reacts strongly with solid-phase $CaCO_3$ as well as with the adsorbed Ca. Since $CaSO_4$ produced from

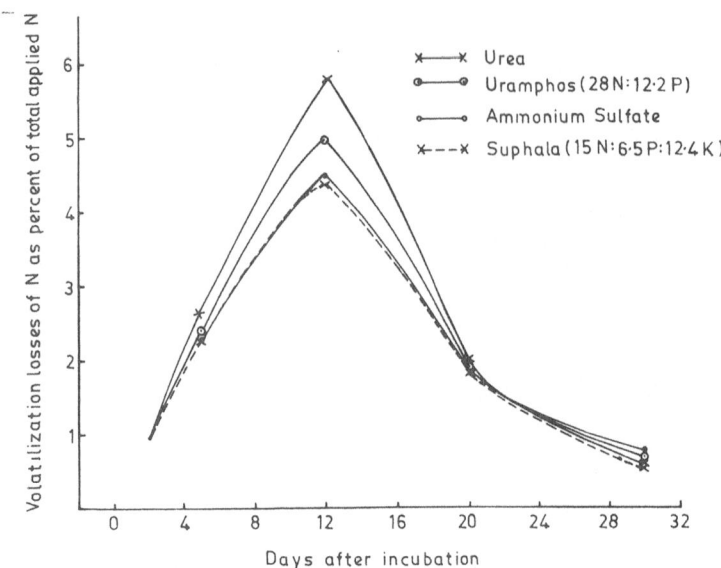

Figure 7. Ammonia loss from vertisols as affected by N source (More *et al.*, 1977).

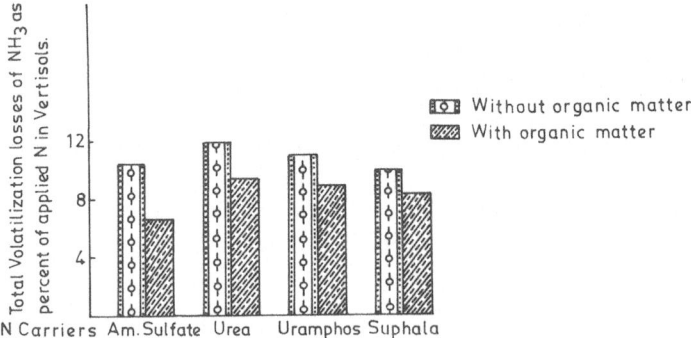

Figure 8. Volatilization loss of ammonia as affected by N source and organic matter (More *et al.*, 1977).

$CaCO_3$ is slightly soluble, the Ca can still act as a replacement cation for NH_4. Therefore, losses of NH_3 with ammonium sulfate are higher than urea in calcareous soils.

The losses through volatilization of ammonia are crucial to agricultural production in India, since urea has emerged as the most important fertilizer, accounting for 73.3% of the fertilizer N consumed in India. Furthermore, popularity of urea is partly due to its high N content (46%), low unit cost, and availability. When urea was applied to a calcareous soil, enzymatic decomposition of urea by urease into ammonium carbonate occurs as follows:

$$CO(NH_2)_2 + 2H_2O \xrightarrow[H_2O]{\text{urease}} (NH_4)_2CO_3 \tag{1}$$

$$(NH_4)_2CO_3 \longrightarrow CO_2 \uparrow + 2NH_3 \uparrow + H_2O \tag{2}$$

The hydrolysis product $(NH_4)_2CO_3$ provides alkalinity for higher losses of ammonia through volatilization over other complex fertilizers. Recent studies by Fenn et al. (1981, 1982) indicated that the NH_3 losses can be effectively controlled by the use of Cl^- and NO_3^- salts of Ca and K with surface-applied urea in calcareous soils.

More et al. (1977) reported that the presence of phosphates in Uramphos and NO_3^- in Suphala reduces volatilization losses of ammonia. The lowest loss of ammonia (10%) from Suphala is due to the presence of NO_3^-. Fenn and Kissel (1974) showed no increase in % N lost with increasing rates of surface application of NH_4NO_3 to a Houston Black Clay with a CEC of 580 mmol/kg^{-1} and $CaCO_3$ content of 29% by weight. Inorganic N compounds where anions of the NH_4 salts are NO_3^-, Cl^-, and H_2PO_4 react with $CaCO_3$ present in vertisols and do not produce insoluble Ca precipitates, resulting in lower NH_3 losses.

A survey of the literature clearly brings out the importance of ammonia volatilization of urea in vertisols with high pH, $CaCO_3$ (%), and low moisture levels. It should be emphasized that there have been few studies on loss of volatilized ammonia under field conditions and these have been hindered by lack of techniques for studying their losses.

The growing acceptance of urea as a major nitrogen source for agricultural production, and the increasing incidence of its use as surface application because of the high energy cost for deep placement has caused a number of researchers to reevaluate the importance of ammonia volatilization as a major avenue of loss of fertilizer nitrogen. The use of acids, urease inhibitors, and addition of Ca, Mg, or K salts with urea for potential control of NH_3 losses proposed by Fenn and co-workers (Fenn and Kissel, 1973; Fenn et al., 1981, 1982) should receive attention in regions where deep placement of nitrogen is impossible or uneconomical for vertisols with high pH, $CaCO_3$ and low moisture levels.

C. Response of Crops to Nitrogen

Vertisols are low in available soil N and generally respond well to fertilizer N application. The magnitude of nitrogen response to vertisols is in the following order: maize (Zea mays L.) > sorghum > pearl millet > wheat. The responses of some field crops to N on farmer's field are shown in Figure 9. The response of wheat to fertilizer N in the farmer's field is about 10 kg grain/kg^{-1} N (Figure 9), which the response in the case of post-rainy

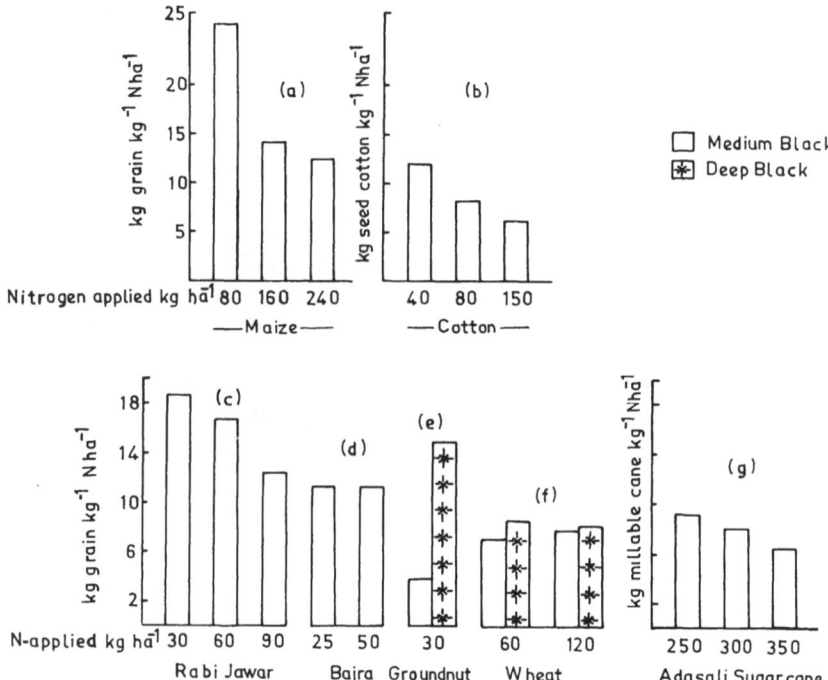

Figure 9. Response of field crops to nitrogen: (a) maize, (b) cotton, (c) rabi jowar, (d) bajra, (e) groundnut, (f) wheat, (g) Adasali sugarcane (Prasad and Subbaiah, 1982).

season sorghum and maize and pearl millet (*Pennisetum americanum* (L.)) is 18, 25, and 12 kg grain/kg^{-1} N/ha^{-1}, respectively. Data on N response to groundnut (*Arachis hypogaea* L.) on deep black soil (Figure 9e) shows a good economic return (Rajendra Prasad and Subbaiah, 1982).

Nitrogen application has certainly found favor in the case of cash crops such as cotton (*Gossypium hirsutum* L.) and sugarcane (*Saccharum officinarum*) (Figure 9b,g). In these crops, the farmers obtain very good returns on investment of fertilizer N.

Based on the results of experiments from farmer's fields, the response of pearl millet and wheat leveled off near 50 and 80 kg N/ha^{-1} in medium-black soils respectively, (Figure 9). Response of post-rainy season sorghum leveled off near 30 kg N/ha^{-1} (Umrani and Bhoi, 1980), and that of maize leveled off at 100 kg N/ha^{-1} (Figure 9).

The responses of sorghum to applied N on the experimental station fields were almost double, with and without similar N fertilizer addition, the response observed on the farmer's fields (Table 17). The differential response to applied N could be attributed to use of improved varieties/

Table 17. Response of dwarf rabi sorghum to nitrogen in the farmer's fields and experimental station, Sholapur

Location	N applied (kg/ha^{-1})	Grain yield (Mg/ha^{-1})	Response $(grain\ kg/N/ha^{-1})$
Farmer's field[a]	0	0.480	—
	25	0.645	6.6
	50	0.785	6.1
Experimental	0	0.880	—
station[b]	30	1.440	18.7
	60	1.880	16.7
	90	1.990	12.3

[a] Mahapatra et al. (1973).
[b] Umrani and Bhoi (1980).

hybrids, weed control to mitigate the adverse effects of weeds on stored water, and control of disease and pests by timely and appropriate plant protection measures.

D. Factors Affecting Crop Response to Nitrogen

Although nitrogen fertilization has been profitable in most crops, its efficiency of utilization has been rather low. Results from experimental stations and from AICARP indicate that some of the prerequisites for N use efficiency in vertisols are (1) responsive genotypes, (2) soil depth and moisture status, (3) crops, season, and cropping systems, and (4) available NO_3 nitrogen status.

1. Use of Fertilizer-Responsive Genotypes

Cultivars differ in their ability to use applied nutrients. As seen from Figure 10, hybrid sorghum (CSH-5) is more responsive to N than other sorghum varieties. Grain yield of CSH-5 on average was 3 Mg/ha^{-1} as compared to the yield of 2 Mg/ha^{-1} obtained with varieties like SPV-346 and SPV-351 (AICSIP, 1984). This establishes the superiority of hybrids over the varieties in their response to N. It is clear from Figure 10 that even if we consider only the major sorghum soils (vertisols) in India, much diversity in yield and N use efficiency exists. Such a diversity within a soil order carries with it the expectations of a variety of soil fertility and plant nutrition problems. However, comparatively little attention has been devoted to these problems, even though India accounts for nearly 31% of the world's sorghum area, with 17.4% of the world's sorghum production (FAO, 1979).

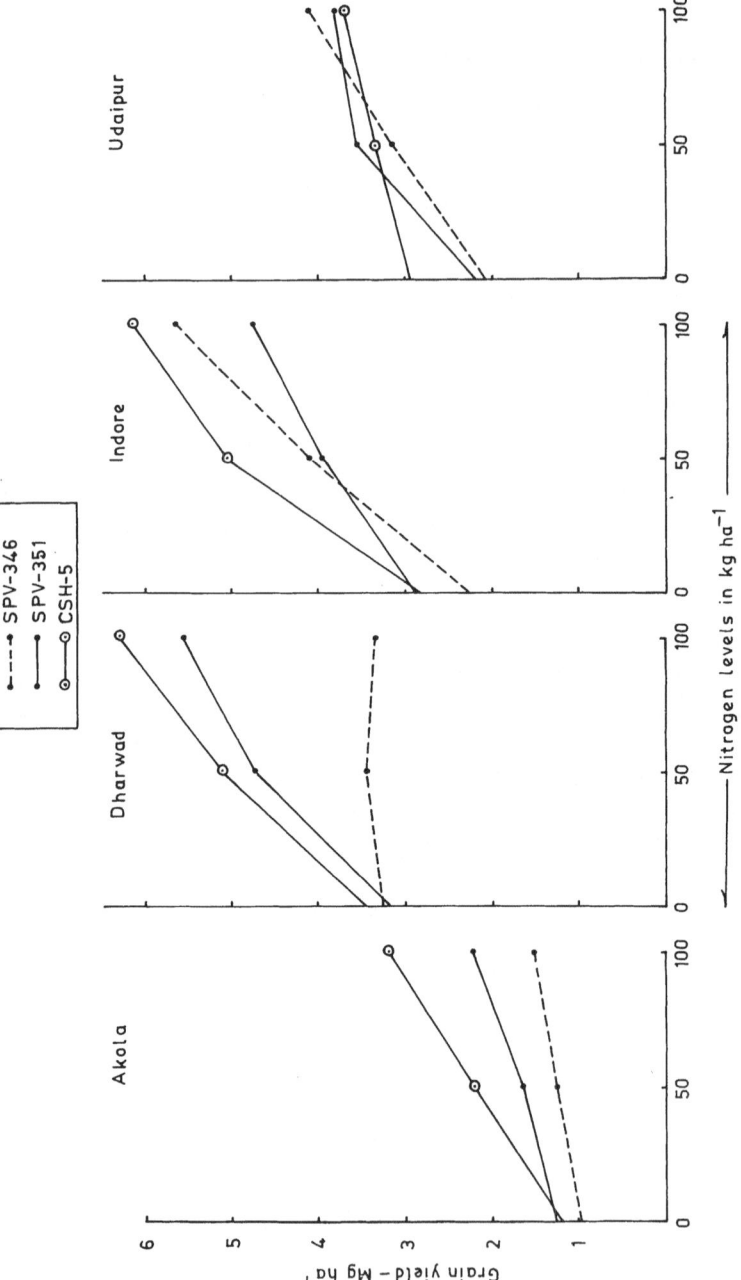

Figure 10. Differential response of sorghum genotypes to nitrogen levels in vertisols at four locations (AICSIP, 1984).

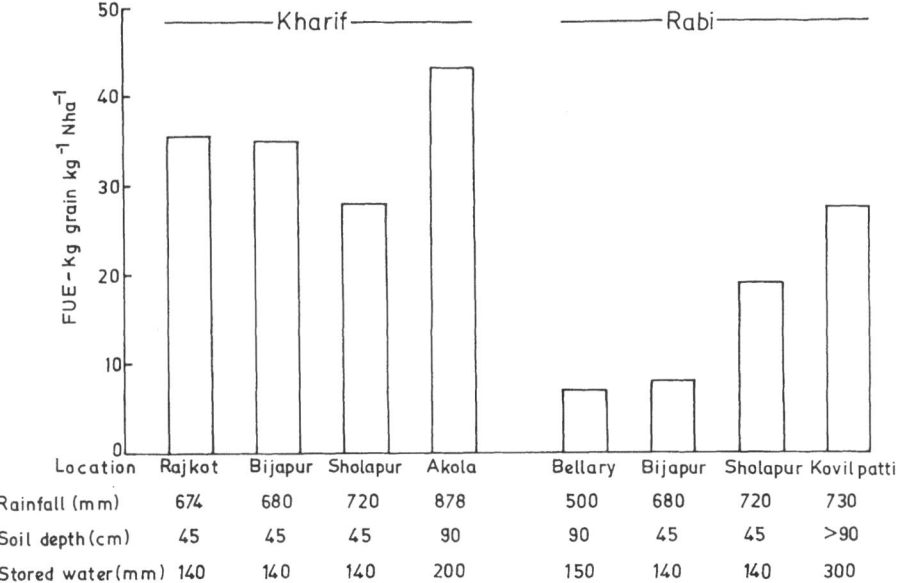

Figure 11. Effect of rainfall, soil depth, and stored moisture on the nitrogen FUE of sorghum in kharif and rabi seasons (Finck and Venkateswarlu, 1982).

2. Soil Depth and Available Soil Water

Sorghum in India is predominantly cultivated in vertisols. Of the 16–18 million hectares grown under sorghum, approximately two-thirds of the area is cultivated during the rainy season and one-third during the post-rainy season. The major point of concern is the ability of the soil profile to store water for crop use. Depending on the soil depth, plant-available water may vary from 100 to over 250 mm. Plant-available water/irrigation mitigates the adverse effects on nitrogen use efficiency, as seen from Figure 11. The results in Figure 11 indicate that the grain yield of post-rainy sorghum increases with the increase in soil depth accompanied by high moisture storage capacity. Further, the work of Umrani and Patil (1983) shows that at a lower level of available water (50 mm), the response to N was up to 25 kg N/ha^{-1} when the soil depth was 300 mm. However, when the available moisture increased to 110 mm at a soil depth of 600 mm, the response was noticed up to 50 kg N/ha^{-1} (Figure 12). Results similar to this were reported by Patil *et al.* (1981) for safflower and pearl millet (Table 18).

3. Cropping Season

Figure 11 shows that the grain yield and fertilizer use efficiency (FUE) increase with increase in rainfall which is mostly received before sowing.

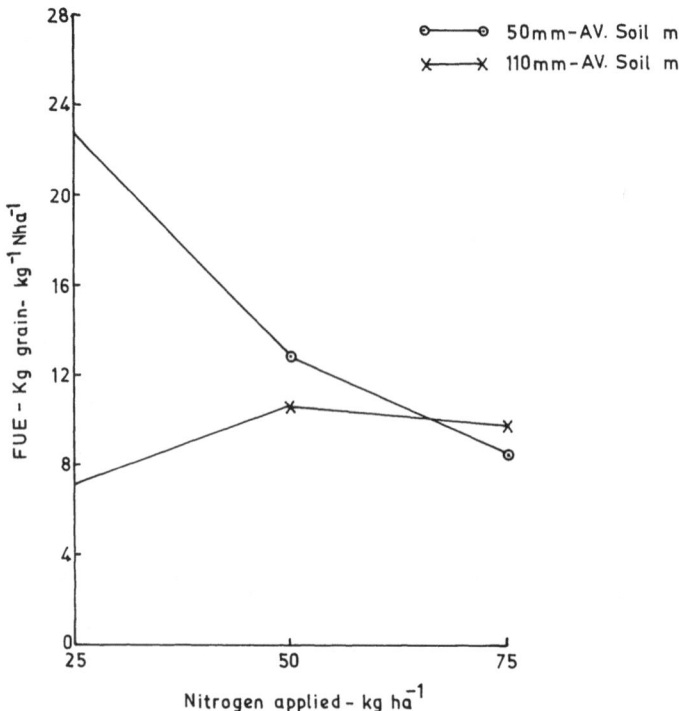

Figure 12. Effect of nitrogen fertilization on FUE at two different levels of soil moisture availability (Umrani and Patil, 1983).

Table 18. Response of safflower and pearl millet to applied N

Level of N (kg/ha^{-1})	Grain yield (Mg/ha^{-1})	kg grain/kg^{-1} N/ha^{-1}
Safflower		
0	0.603	—
25	0.997	15.76
50	1.100	9.94
75	1.401	10.64
100	1.410	8.07
Pearl millet (HB-3)		
0	1.078	—
25	1.601	20.92
50	2.055	19.54
75	2.347	16.92
100	2.573	14.95

Source: Patil *et al.* (1981).

FUE for kharif sorghum ranged from 28 to 43 kg grain/kg^{-1} N/ha^{-1} in contrast to 6.5–27.7 kg grain/kg^{-1} N/ha^{-1} by sorghum raised as a post-rainy season crop. As seen from Figure 11, rabi sorghum is less responsive for the following reasons (Kanwar and Rego, 1983; Finck and Venkates-warlu, 1982): (1) use of less N-responsive varieties; (2) accumulation of nitrates due to mineralization of soil N during the fallow period; and (3) difficulties in placing the fertilizer in moist soil layers to facilitate rapid utilization of the nutrient N.

The replacement of traditional varieties with fertilizer-responsive varieties and deep placement of fertilizers and/or irrigation at sowing time to facilitate crop establishment and better utilization of fertilizers makes a considerable difference in responsiveness of post-rainy season crops to applied N.

4. Cropping Systems

Umrani and Patil (1983) advocate a systems approach for better fertilizer use. For post-rainy sorghum, legume incorporation as a preceding crop helps to utilize the applied fertilizer economically. Further, it is observed by Umrani and Patil (1983) that if the preceding legume crop is fertilized by a small dose of fertilizer (12.5 kg N + 10.9 kg P/ha^{-1}), it helps to mini-mize the nitrogen requirement of rabi sorghum by nearly 25 kg N/ha^{-1}.

Grain yield and FUE-N from a sorghum-safflower double-cropping

Table 19. Effect of nitrogen fertilization on sorghum-safflower double cropping (sorghum hybrid CSH-6 and safflower variety Manjira; 20 kg P/ha^{-1})

Nitrogen applied (kg/ha^{-1})		Grain yield (Mg/ha^{-1})		Total
A	B	Sorghum (A)	Safflower (B)	sorghum + safflower
0	0	3.722	0.628	4.350
60	0	4.885 (19.4)	0.721	5.606 (20.9)
90	0	5.290 (17.4)	0.861 (—)	6.151 (20.0)
30	30	4.710 (32.9)	0.996 (12.3)	5.706 (22.6)
60	30	5.183 (24.3)	1.048 (14.0)	6.231 (20.9)
	CD 5%	0.519	0.266	

Source: Kanwar *et al.* (1984).

Figures in parentheses indicate FUE.

A, Rainy season; B, post-rainy season.

Table 20. Effect of N and irrigation on yields, nutrient uptake, and WUE by rabi sorghum in a stored moisture environment

N applied (kg/ha^{-1})	Number of irrigations	Grain yield (Mg/ha^{-1})	N uptake (kg/ha^{-1})	Fertilizer N recovery (%)	Consumptive water use (mm)	WUE-C (kg/mm^{-1}/ha^{-1})	WUE-I (kg/mm^{-1}/ha^{-1})	FUE
0	0	3.458	75.0	—	114.5	3.02	—	—
	2	4.035	105.5	—	268.5	1.50	4.4	—
80	0	4.053	102.6	34.5	150.5	2.81	—	0.74
	2	5.319	164.8	74.1	335.0	1.68	9.7	1.61

Source: Sardar Singh and N. Seetharama quoted by Kanwar *et al.* (1984).

experiment show that the nitrogen use is slightly higher in the double-cropping system than under single cropping (Table 19). Further, if the N is split and applied to both rainy season and post-rainy season crops, then the FUE-N is enhanced. Nitrogen use efficiency for the rainy season sorghum alone is highest under low rate of 30 kg N/ha^{-1} (Kanwar *et al.*, 1984).

Further, Kanwar *et al.* (1984) observed a twofold increase in WUE-I with 80 kg N/ha^{-1} as compared to no nitrogen application. Irrigation was found to increase the recovery of applied N from 34.5% to 74.1% (Table 20).

5. Available N (NO$_3$-N) Status and Irrigation

The data shown in Table 21 clearly indicate that the responsiveness of a crop variety to applied N is largely governed by the available N and moisture status in the soil (Kanwar and Rego, 1983). The response of rabi sorghum to 80 kg N/ha^{-1} was 0.60 Mg/ha^{-1} when the available NO$_3$-N was 40 kg/ha^{-1} under no irrigation. After an exhaustive maize crop, the response to same dose of N increased to 1.424 Mg/ha^{-1} with no irrigation. Two irrigations increased the difference in response to N by 1.283 Mg/ha^{-1} and 3.449 Mg/ha^{-1} when the available NO$_3$-N levels were 40 and 13 kg/ha^{-1}, respectively. This apparently suggests higher response to added N when available NO$_3$-N was low. Further, the effect of two irrigations under adequate N fertilization on sorghum yield responses were 1.265 and 2.482 Mg/ha^{-1} where initial soil NO$_3$-N levels were 40 and 13 kg/ha^{-1}, respectively.

Table 21. Nitrogen status and irrigations on post-rainy season sorghum (CSH-8R) yield on a vertisol at ICRISAT

Nitrogen applied (kg/ha^{-1})	Number of irrigations after emergence	Cropping season	Available NO$_3$-N (kg/ha^{-1})	Grain yield (Mg/ha^{-1})
0	0	Rainy season	40	3.458
	1	fallow		3.771
	2			4.035
80	0			4.053
	1			4.583
	2			5.318
0	0	Rainy season	13	1.083
	1	maize		1.506
	2			1.540
80	0			2.507
	1			4.158
	2			4.989

Source: Kanwar and Rego (1983).

Table 22. Influence of rate and method of application of urea N on yield of dry matter and N uptake by Kharif sorghum during 1981

N applied (kg/ha^{-1})	Application method	Grain yield (Mg/ha^{-1})	N uptake (kg/ha^{-1})
0		2.72	38.6
37	Split band	3.61	52.9
74	Split band	5.22	84.4
111	Split band	5.33	81.9
148	Split band	5.31	88.6
74	Surface	4.26	62.1
74	Incorporation	4.11	60.2
SEM*		0.23	4.1
F value		***	**

Source: Moraghan *et al.* (1984); reprinted with permission.
**P = 0.01; ***P = 0.001.

Table 23. Effect of method of application on the fate of labeled urea N applied to sorghum at the rate of 74 kg N/ha^{-1}

Application method	^{15}N recovery %					N loss %
	Soil	Grain	Stover	Chaff	Total	
Split band	38.6	37.7	14.8	3.1	94.2	5.8
Surface	41.8	20.9	7.7	1.9	73.3	27.7
Incorporation	45.2	20.0	6.9	2.0	74.1	25.9
SEM	2.7	1.4	8.7	0.1	2.1	
F value	NS	***	***	***	***	

Source: Moraghan *et al.* (1984); reprinted with permission.
NS, not significant.
***P = 0.0001.

E. Method of Application

Nitrogen use efficiency in vertisols can be considerably enhanced by proper application methods. Incorporation, split application, and banding of N fertilizers below the soil surface reduce volatilization and delay nitrification.

Recently, Moraghan *et al.* (1984) conducted field studies on vertisols to determine the effect of different methods of ^{15}N fertilizer application on sorghum grain yield and its recovery. The data in Table 22 indicate that split band (SB) application of 74 kg N/ha^{-1} resulted in 22% and 27% increase in grain yield and 37% and 36% increase in total N uptake over surface and incorporation methods of N application, respectively.

Recovery of labeled urea N was greatest with SB application method

Table 24. Effect of rate and method of application urea N applied to Kharif sorghum on yield of safflower, plant uptake, and recovery of N

N rate (kg/ha^{-1})	Application method	Grain (Mg/ha^{-1})	Total plant N (kg/ha^{-1})	[15]N recovery (%)
0		0.78	23	—
74	Split band	0.90	27	1.5
74	Surface	1.05	32	4.2
74	Incorporation	9.70	30	3.8
SEM		0.10	3	1.1
F value		NS	NS	NS

Source: Moraghan *et al.* (1984); reprinted with permission.
NS, not significant.

(Table 23). The combined soil-plant uptake data show that over 94% of the urea N applied was recovered with SB treatment. In contrast, under surface and incorporation methods, the recoveries were only 72%, 74%, respectively.

Under rain-fed farming at Hagari, Karnataka, use of seed-fertilizer drill gave highest sorghum yield of 0.58 Mg/ha^{-1}, representing an increase of 18.7% over control (Gopalakrishna Rao *et al.*, 1975).

F. Residual Effect of N

The data in Table 24 obtained by Moraghan *et al.* (1984) show that only a small part of the residual labeled N applied to the rainy season sorghum was recovered by the safflower in the post-rainy season. The fate of this immobilized fertilizer N is not known. Thus, studies are needed to improve the response as well as recovery of applied nitrogen and to determine the fate of immobilized N by the soil. The limited availability of the residual fertilizer N to the subsequent crops during the post-rainy season as explained by these authors could be due to limited soil moisture in the surface layer and transformation to organic N, a portion of which gets immobilized with sorghum roots. The retention of significant quantities of fertilizer N in vertisols low in total N raises the unanswered question of the consequences of long-term N fertilization associated with introduction of improved cropping systems and needs detailed study (Moraghan *et al.*, 1984).

VII. Phosphorus

Phosphorus is considered a major constraint for crop production in vertisols. These soils have extremely high capacities to immobilize phosphorus, and thus its deficiency has become widespread.

Crop recovery of added phosphates seldom exceeds 20%, and this calls for ways and means to increase the efficiency of P utilization by crops. The purpose of this review is to summarize the present knowledge of the forms of P and its management in Vertisols.

A. Phosphorus Forms in Vertisols

The total phosphorus content of vertisols of India ranges from 9.8 to 16 g/mg^{-1}. In vertisols of heavy rainfall area of Gujarat, total P values were 16 and 9 g/mg^{-1} under fruit crops and teak forest, respectively (Mehta et al., 1979). More et al. (1979) reported that total phosphorus of vertisols of Marathawada at depths of 0–225 mm vary from 722 to 10,574 mg/kg^{-1}. There was a significant negative correlation of total phosphorus with $CaCO_3$, but pH, organic carbon, and clay content did not bear any relationship. The total phosphorus was highly correlated with available phosphorus ($r = 0.62$). The results reported by More et al. (1979) clearly suggest that for effective recommendation of phosphorus application to vertisols, information on $CaCO_3$ content should be considered along with soil test value for available P.

Mehta et al. (1979) reported that available P determined by the method of Olsen et al. (1954) in the deep black soils of Gujarat, India, varies from 2 to 34 kg/ha^{-1}. The available P was maximum in the surface layers and decreased with depth. Positive and significant correlations between available P content and organic matter ($r = 0.4827$) and organic P ($r = 0.4844$) were reported. A significant negative correlation was reported between available P vs. clay and sesquioxides ($r = -0.4876$).

The available P content of vertisols of Marathawada ranged from 4 to 27 kg P/ha^{-1}. Available phosphorus was negatively correlated with $CaCO_3$ contents ($r = -0.59$). This suggests surface adsorption of phosphorus on $CaCO_3$ results in low availability of P in calcareous soils (More et al., 1979).

The distribution of different forms of P in some vertisols of India is summarized in Table 25. The solid inorganic forms of phosphorus are usually divided into three active fractions and two relatively inactive fractions. The active fractions can be grouped into calcium-bonded phosphates (Ca-P), aluminium-bonded phosphates (Al-P), and iron-bonded phosphates (Fe-P). Calcium phosphates are present as films or as discrete particles, whereas Al-P and Fe-P occur as films and/or are simply adsorbed on clay or silt surfaces. The relatively inactive fractions are the occluded and reductant-soluble forms. Occluded phosphorus consists of Fe-P and Al-P surrounded by an inert coat of another material that prevents the reaction of these phosphates with the soil solution. Reductant soluble forms are covered with an inert material that may be partially or totally dissolved under anaerobic conditions. This form of P is of importance in the P nutrition of rice under submerged conditions.

Table 25. Phosphorus fractions in some vertisols

Location	Forms of P (mg/kg^{-1})							
	Saloid P	Al-P	Fe-P	Ca-P	R-P	Occl-P	Total P	Organic P
Tamilnadu[a]	5.0	43.0	8.0	78.0	30.0	5.0	486.0	237.0
Hyderabad[b] (ICRISAT)	—	17–35	21–41	42–93	—	—	—	—
Karnataka[c]	1.9	36.7	9.6	76.4	35.7	11.4	392.0	220.8
Madhya Pradesh		11.0	21.3	166.4	—	—	—	—
Nandyal (A.P.)		17.0	21.0	70.0	41.0	—	—	—

[a] Kothandaraman and Krishnamoorthy (1979).

[b] Sahrawat (1977).

[c] Doddamani (1982).

The forms of inorganic phosphorus present in vertisol are presented in Table 25. In vertisols, calcium phosphates and the reductant soluble phosphates are the dominant inorganic forms of phosphorus. This suggests that vertisols are not highly weathered.

B. Factors Affecting P Content in Vertisols

Transformation of native and applied P in soils is greatly influenced by crop, fertilizer, and management practices. There is lack of data on phosphate transformations as affected by soil and cropping systems.

Mehta *et al.* (1979) observed a great variation in the forms of inorganic phosphorus in different cropping systems in vertisols. Vertisols in heavy rainfall areas of Gujarat cropped to fruit trees and teak forest contained 1131 and 795 mg/kg^{-1} P, respectively, as compared to soils under grasses (40 mg/kg^{-1}) and cereals (238 mg/kg^{-1}).

Singhania and Goswami (1978) studied the transformation of native and applied P in a rice-wheat cropping sequence. The data shown in Table 26 clearly indicate that P transformations in vertisols were greatly affected by cropping. Native Al-P and reductant P (RP) decreased by 21.6% in vertisols where rice was grown as compared to uncropped conditions. This indicated that the rice crop used native Al-P and RP as sources of P. A definite decrease in native Fe-P and RP fractions when wheat was grown following rice with not much change in Al-P and Ca-P fractions indicated that Fe-P and RP fractions contributed more toward the requirement of wheat (Table 26).

C. Crop Response to P in Vertisols

Considerable research has been carried out on the response of field crops to applied P on vertisols occurring in different parts of India. (AICARP,

Table 26. Effect of cropping on native phosphorus fractions (no phosphorus treatment) in vertisols

	Phosphorus fractions (mg/kg^{-1})			
	Al-P	Fe-P	R-P	Ca-P
No crop	14.3[a]	21.7	57.0	74.5
	15.5[b]	23.9	52.1	80.0
After rice	11.2	26.8	44.7	77.4
After rice and wheat	10.9	20.5	41.5	75.0

Source: Singhania and Goswami (1978); reprinted with permission.

[a] Corresponding to the period of rice crop (122 days).

[b] Corresponding to the period of both rice and wheat (227 days).

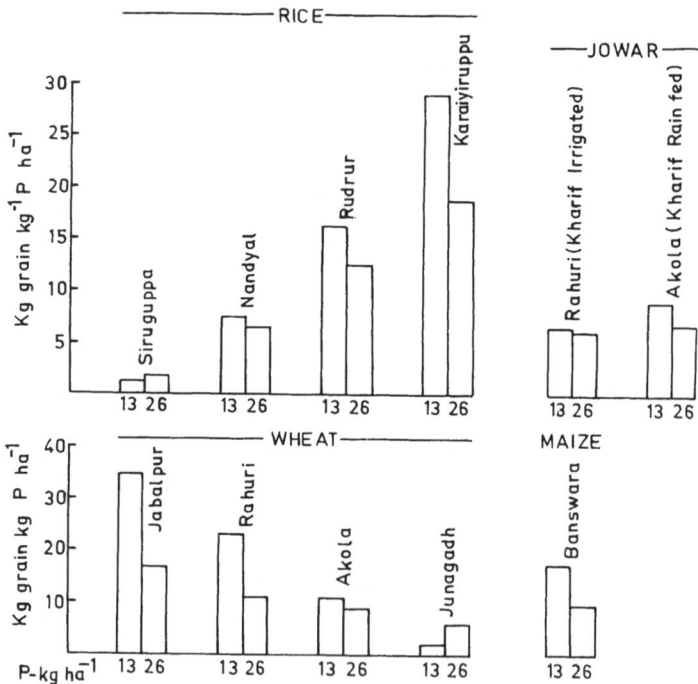

Figure 13. Response of rice, jowar, wheat, and maize to P in vertisols (AICARP, 1981).

Table 27. Effect of phosphate application on
post-rainy season sorghum

P (kg/ha^{-1})	Grain yield (Mg/ha^{-1})
0	1.27
22	1.05
44	1.15
66	1.24
SE ±	0.67
CD (5%)	Not significant

Source: Patil *et al.* (1981).

1974–1980). The responses of crops like wheat, rice, maize, and sorghum to applied P are shown in Figure 13. The data in Figure 13 indicate that maximum yield could be obtained by applying 13 kg P/ha^{-1}. The response was found to range from 2 to 34, 1 to 29, 10 to 19, and 6 to 9 kg grain/kg^{-1} P/ha^{-1} in case of wheat, rice, maize, and sorghum, respectively.

D. Factors Affecting Crop Response to P

1. Initial Available P Status

Results from a field trial on medium black soil at Akola with safflower as a test crop indicate a positive response to P in soils containing less than 5.2 kg available P/ha^{-1} (Hanwante *et al.*, 1981). Results similar to this are available from studies carried out by Patil *et al.* (1981) on the phosphate requirement of sorghum variety M 35-1. Pooled grain yield data (1973–76) presented in Table 27 indicate lack of response to phosphate. The authors attribute the lack of response to the level of available phosphate in the soil. The soil on which the crop was grown contained 8.7 kg P/ha^{-1}. Second, the requirement of phosphate for M 35-1 is very low (11–12 kg P/ha^{-1}), which the plant could obtain from native phosphate alone. The studies clearly illustrate that a positive response to applied P is possible when applied to soils testing low to medium levels of available P and crops having higher P requirements.

2. P-Fixing Capacity

The data in Figure 13 show that the average response in case of rice was 22 kg grain/kg^{-1} P/ha^{-1} in low P-fixing vertisols of Karaiyiruppu as compared to 1.5 kg grain/kg^{-1} P/ha^{-1} in high P-fixing soils of Siruguppa. Similarly for wheat, the response was 34 and 17 kg grain/kg^{-1} P/ha^{-1} in low and high P-fixing vertisols of Jabalpur and Akola, respectively. The vertisol from Junagadh is a low P-fixer, but still the response of wheat is 2–5 kg grain/ kg^{-1} P/ha^{-1} partly because of inadequate moisture in the profile, and also

Table 28. Combined influence of irrigation and phosphorus on available P in surface soil, haulm yield, and P uptake by haulm (kg/ha^{-1}) at harvest in groundnut in vertisols

Treatment (P kg/ha^{-1})	Available P			Haulm yield			P uptake by haulm		
	I_{25}[a]	I_{50}	I_{75}	I_{25}	I_{50}	I_{75}	I_{25}	I_{50}	I_{75}
0	13.1	14.4	15.7	2674	3442	4003	10.0	10.5	11.9
10.9	14.9	17.0	17.8	2813	3462	4265	11.0	12.2	13.7
21.8	14.0	20.1	23.2	3073	3878	4531	12.5	14.4	15.1
32.7	16.8	20.4	26.7	2878	3792	4781	10.2	13.8	16.3
CD (5%)		6.1			221			1.4	

Source: Zalawadia and Patel (1983).

[a] I_{25}, I_{50}, and I_{75} represent 25%, 50%, and 75% available soil moisture.

because the shallow root system of wheat prevented tapping of subsoil moisture for better utilization of applied P.

Field studies by Zalawadia and Patel (1983) with groundnut as a test crop revealed response to P under adequate moisture conditions. Results shown in Table 28 indicate that irrigation to attain 75% of available moisture in conjunction with 33 kg P/ha^{-1} significantly outyielded other treatments with respect to haulm yield, available P in the soil, and P uptake by haulm.

3. Dose of Nitrogen

Field experiments on the response of rice, maize, and wheat to N and P on vertisols carried out by AICARP reveal the strong synergistic interaction of these two nutrients in soil (Figure 14). Responses to P have been quite marked when P was applied in conjunction with nitrogenous fertilizers.

A comparison of the yield data in Figure 14 for rice at N_0P_{26} with N_0P_0 show that the increase is only 5% and 11% at Karjat and Karaiyiruppu, respectively. But when 26 kg P/ha^{-1} was applied in conjunction with 60 kg N, the percent increases in yields over N_0P_0 were 43% at Karjat and 93% at Karaiyiruppu, suggesting better response of crops to P in the presence of N. The yield differences at these two locations could be due to the differences in the available moisture status, since the rice crop at Karjat was raised as a rain-fed crop, and at Karaiyiruppu it was raised under irrigation. Under submergence, the reductant soluble P gets dissolved and thereby increases the concentration of soluble P, which results in increased P uptake by the crop leading to higher yields at Karaiyiruppu.

In the case of wheat and maize, the increase in yields at N_0P_{13} over control were found to range from nil to 40%, respectively. At $N_{120}P_{26}$, the increase in maize yield was 194% and that of wheat was 272% over N_0P_0 (Figure 14). Grain yield data suggest the possible utilization of P from the

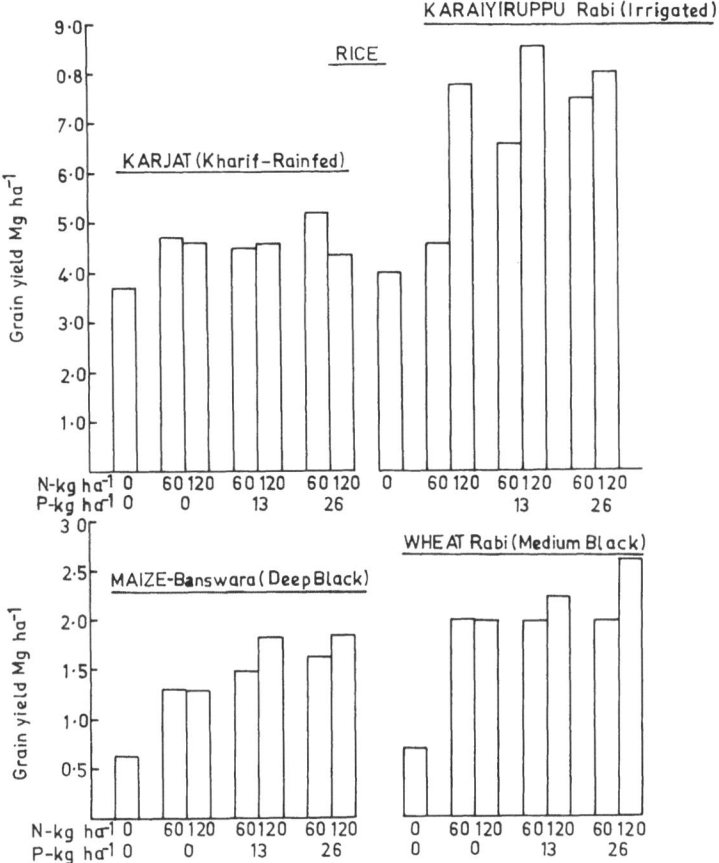

Figure 14. Response of cereals to N and P in vertisols (AICARP, 1981).

relative inactive occluded Fe-P, Al-P, and reductant-soluble phosphates in the presence of N.

4. Cropping Systems

Phosphorus being a costly input, it is necessary to work out the phosphate requirement for the most important cropping systems followed in various agroclimatic zones in the country. Attempts to elucidate the crop characteristics and the cropping sequence for matching the fertilizer needs of crop with a view to economizing and rationalizing P application are needed.

Results of a 3-year study carried out by the AICARP are presented in Table 29. For continuous rice rotation, application of 26 kg P/ha^{-1} is inadequate to meet the heavy demands of rice which removes roughly the same amount of phosphate per ton of grain production. In the rice-wheat

Table 29. Responses to P application in Kharif or rabi in some cropping systems in black soils (Mg/ha^{-1})

Cropping sequence	Location	Total rotational response to applied P (26 kg/ha^{-1})		
		A	B	C
Rice-rice	Maruteru	0.42	0.30	0.34
	Karaiyiruppu	0.34	0.32	0.37
Rice-wheat	Jabalpur	1.99	1.88	2.34
Sorghum-wheat	Indore	1.88	1.90	2.24
	Siruguppa	2.55	2.88	2.94

Source: Goswami and Mohinder Singh (1976).

A, Rainy season; B, post-rainy season; C, rainy and post-rainy season.

rotation, the decrease in yield suggests that 26 kg P/ha^{-1} is inadequate to meet the full requirement of wheat and rice. This clearly indicates that both rice and wheat should be fertilized at an optimum level (Goswami and Mohinder Singh, 1976). From the results of the sorghum-wheat sequence, it is difficult to draw definite conclusions. However, the results suggest that application of P to wheat is economical as against its application to both the crops (Table 29).

5. Soil Solution P

Vertisols are known to have high P-fixation capacity, ranging from 304 to 417 mg/kg^{-1} of soil, and seldom have more than 0.01 mg/L^{-1} of P in soil solution (ACBSR, 1984), in contract to 0.2–5 mgL^{-1} of P to produce maximum yields (Barber, 1983). Moreover, the concentration of P in soil solution is vital, since plants growing in soil absorb P from the soil solution at the root surface by diffusion. Furthermore, soil solution must contain sufficient P to provide concentration gradient necessary for net movement to the root. This would possibly explain the lack of response to P in vertisols, since the amount of P added seldom exceeds 26–23 kg P/ha^{-1}, which is quite inadequate to meet the P-fixing capacity of the soil and the crop requirement. Relatively little work on the relationship between the amounts of inorganic P to be added to the soil to arrive at a desired level of soil solution P for maximum yields has been done.

Results of recent work on the amounts of P needed to attain a desired level of solution P are shown in Table 30 (ACBSR, 1984). Data shown in Table 30 show that critical P levels for vertisols vary with parent material from which they have been derived. To maintain a critical concentration of 0.2 mg/L^{-1} of P optimum for field crops, 8–22 mg P has to be added per kilogram of soil. This approach provides a method for studying reactions of

Table 30. Phosphorus fixation and amount of P needed to obtain 0.2 Mg/kg^{-1} in soil solution of some vertisols of diverse origins

Parent material	Percent clay	Fixed P (mg/kg^{-1})	
		Adsorption maxima	At 0.2 mg in soil solution
Basalt	61	341	19
Granite	65	312	08
Granite gneiss	59	417	08
Limestone	72	304	22
Schist	44	356	16

Source: ACBSR (1984).

P fertilizers that is more closely related to plant needs than some of the traditional methods now in use.

E. Direct, Residual, and Cumulative Effects of P Fertilizer in Cropping Systems

Vertisols contain appreciable reserves of phosphorus, but it is not released for the crop in adequate amounts mainly owing to its presence as insoluble phosphates. This calls for phosphorus application in amounts high enough to compensate for crop removal and fixation by the soil. A heavy rate of P application for the main crop will have residual benefit for the succeeding crop, since phosphorus removal by the first crop rarely exceeds 10–20% of added phosphorus.

No definite conclusion can be drawn from the results presented in Table 31 on the direct, residual, and cumulative responses of fixed, 1-year, two-cereal crop rotations to the application of 26 kg P/ha^{-1}. The amount of P applied is inadequate to meet the requirements of the crop, generally inadequate to compensate for fixation and crop needs and also the fixation capacity of the soil (AICARP, 1975).

However, when fertilization of the cropping system is being considered, it is important to apply the amount of nutrients that give the greatest economic return. The place to apply the fertilizer in the cropping system will depend on the response of each crop. Often, it is most profitable to apply much of the phosphate to the highest-value crop and allow the lower-value crops to obtain their phosphorus from the residual phosphorus remaining in the soil (Barber, 1983).

F. Sources of P

The most important characteristic of phosphatic fertilizer that affects the response or uptake by crops is its water solubility. Phosphatic fertilizers

Table 31. Direct, residual, and cumulative responses (over no P) of fixed, 1-year, two-cereal crop rotations to the application of phosphorus at 26 kg P/ha^{-1} at Model Agronomic Centers (pooled for 3 years)

Center	Soil group	Rainy season			Post-rainy season		
		Direct	Residual	Cumulative	Direct	Residual	Cumulative
Rice-rice (Mg/ha^{-1})							
Maruteru	Black	0.131	0.163	0.121	0.143	0.288	0.233
Karaiyiruppu	Black	0.274	0.121	0.192	0.199	0.070	0.176
Rice-wheat (Mg/ha^{-1})							
Jabalpur	Black	1.219	0.938	1.327	0.944	0.773	1.017
Jowar-wheat (Mg/ha^{-1})							
Indore	Black	1.481	1.136	1.506	0.764	0.403	0.734
Siruguppa	Black	1.644	1.796	1.895	1.088	0.904	1.051

Source: AICARP 1972–73 (1974) and 1973–74 (1975), and unpublished data for 1974–75.

Table 32. Percent utilization of different P fertilizers by rice and wheat

	Crop	
P source		Rice/Wheat
Superphosphate	1.4	23.9
Monocalcium phosphate	3.9	17.4
Dicalcium phosphate	0.6	12.1
Calcium metaphosphate	2.4	19.7
Monoammonium phosphate	3.4	28.4
Diammonium phosphate	2.2	17.1
NH_4 aged superphosphate	2.3	1.9

Source: Datta and Venkateswarlu (1968).

Table 33. Effect of P application to the soil and to the roots in the form of slurry on the grain yield of rice during wet season

Method of application	kg/ha^{-1}	Grain yield (Mg/ha^{-1})
Soil	0	3.7
	13	4.1
	26	4.3
Slurry to roots	0	3.6
	4	3.8
	8	4.0
	12	4.2
	16	4.1
LSD P = 0.05		0.4

Source: AICRIP (1975).

vary in their water solubility from 0% to 100%, which has a bearing on their agronomic effectiveness. Chemical composition of the fertilizer, granule size, and method of application influence markedly the response to P fertilizers.

Datta and Venkateswaralu (1968), using ^{32}P as tracer, evaluated the phosphates of varying solubilities in a greenhouse. The results in Table 32 indicated the superiority of monocalcium phosphate and monoammonium phosphate for rice and wheat, respectively, grown in black soils.

G. Method of Application

It has been reported (AICRIP, 1975) that the problem of high phosphorus fixation in rice soils can be alleviated by dipping rice roots in superphosphate–cow dung slurry (Table 33).

VIII. Potassium

Next to nitrogen and phosphorus, potassium is a limiting fertilizer element
in Indian soils. Potassium removal by crops equals, and in many cases exceeds, that of nitrogen. However, research work on potassium dynamics in
soil, potassium status, and response of different crops to potassium under
different soil groups is lagging behind because of the general impression
that most Indian soils are well supplied with this element. With the introduction of high-yielding varieties/hybrids in recent years, it has become
increasingly important to replenish the soil with all nutrients, including
potassium. In this review, an attempt is made to present the available information on the above aspects and areas that require further research are
also indicated.

A. Origin of Soil Potassium

Most of the potassium in soils is known to be present within the layer
silicates. The important K-bearing minerals are micas and feldspars.
Chemically, soil potassium is usually divided into three categories—water
soluble, exchangeable, and nonexchangeable. The various forms are in
equilibrium with one another and are a function of the mineralogical
makeup of the soil.

B. Distribution and Forms of Potassium

The total content of K in vertisols of India is about 1%. Data on the distribution and forms of potassium in vertisols are presented in Table 34.
Water-soluble potassium in vertisols ranges from traces to 9 mmol/kg^{-1}

Table 34. Forms and status of potassium in vertisols of India (in mmol/kg^{-1})

State	Water-soluble	Exchangeable	1 N HNO$_3$-soluble	HCl-soluble	Total
Andhra Pradesh	0.1	5.5	19.7	—	—
Bihar	Trace	2.5	36.0	—	231.9
Gujarat	0.9	12.8	81.0	91.9	—
Karnataka	Trace	5.5	15.0	—	173.4
Madhya Pradesh	Trace	10.0	—	—	—
Maharashtra	0.1	4.1	19.9	62.6	139.1
Rajasthan	0.1	10.0	36.0	120.0	256.0
Tamilnadu	0.3	5.1	21.0	7.2	230.0
Uttar Pradesh	Trace	5.6	31.0	121.5	600.0

Source: Zende (1978).

soil constituting 1.5% of exchangeable K. Exchangeable potassium content varies from 2.5 to 12.8 mmol/kg^{-1}. This represents 2.5% and 21% of total and K soluble in 1 N HNO_3, respectively. Exchangeable potassium in vertisols of Gujarat (Mehta, 1976), Madhya Pradesh, and Rajasthan is approximately twice that found in vertisols of Andhra Pradesh, Karnataka, Maharashtra, Tamilnadu, and Uttar Pradesh (Verma and Verma, 1968; Godse and Gopalakrishnappa, 1976; Kadrekar, 1976; Krishnamoorthy et al., 1976; Mishra et al., 1970). This implies that the source of K in vertisols of Gujarat, Madhya Pradesh, and Rajasthan is from micas of trioctahedral origin.

Potassium soluble in 1 N nitric acid in vertisols was found to vary from 15 to 36 mmol/kg^{-1}, representing 26.5% of total K. Vertisol of Gujarat (Mehta, 1976) has 8 mmol of HNO_3 soluble K and is approximately three times that found in vertisols from other states.

C. Factors Affecting the Forms of Potassium

1. Parent Rock

In a study on K status of Tamilnadu, Krishnamoorthy et al. (1976) report that black soils that originate from traps and hornblende are low in potassium because of the presence of lime and sodalime feldspars. Kalbande and Swamynatha (1976) presented experimental results that demonstrate a significant relationship between different forms of potassium in vertisols and the parent material from which the soil has been derived (Table 35). Vertisols developed from granitized schist is rich in water-soluble (12.5 mg/kg^{-1}), exchangeable (272 mg/kg^{-1}), fixed (877 mg/kg^{-1}), and total (7800 mg/kg^{-1}) potassium in contrast to vertisols derived from hornblende granulite, calcic gneiss, amphibolite, and chlorite schists. The data on total K_2O and the potassium extracted with boiling HNO_3 indicate that the vertisols of Tungabhadra catchment area contain nearly 10% mica and contribute to the K status in these soils as reported by Kalbande and Swamynatha (1976).

Table 35. Forms of K in vertisols of diverse parentage (in mg/kg^{-1})

Parent material	Water-soluble	Exchangeable	Fixed	Total
Hornblende-granulite	5.7	188.0	550	6770
Calcic gneiss	8.6	195.0	467	6770
Granitized schist	12.5	272.0	877	7800
Amphibolite schist	5.7	217.0	767	5560
Chlorite schist	9.0	213.0	682	6910

Source: Kalbande and Swamynatha (1976).

2. Particle Size

The relationship of different forms of K to particle size is not the same in all vertisols, but the correlation between K fractions and increasing size of soil particles is significant and negative. The clay fraction of coarse-textured vertisol from Rajasthan and Uttar Pradesh contributed 53%, 65%, and 69% to total K-, HCl-, and HNO_3-soluble fractions of K, respectively (Mehrotra and Gulabsingh, 1970), the only exception being the clay fraction of vertisol from Uttar Pradesh, which contributed 58% to HNO_3-soluble K. Lodha and Seth (1970), studying the relationship between forms of K to particle size in vertisols of Rajasthan, concluded that neither the clays nor the soils are rich in K-bearing minerals, and from the total potassium values, it was inferred that the clays were kaolinitic or montmorillonite types. In vertisols of Rajasthan and Uttar Pradesh, the amount of K in HCl-soluble fraction represents nearly 70% and 40% of total potassium, respectively (Lodha and Seth, 1970; Mehrotra et al., 1973). This suggests that a greater proportion of K is held in dioctahedral clay mica and feldspars.

D. Crop Response to Potassium

Responses to potassium application in black soils have been inconsistent, though the removal of potassium by crops in most cases far exceeds that of nitrogen. Lack of response to applied K could possibly be due to (1) differences in fixation and release of potassium from soil through natural process, or (2) potassium application rates being generally inadequate to compensate for fixation and crop needs.

Zende (1978) observed yield responses to potassium ranging from 17.4% to 51% with sorghum as a test crop. Further, he observed a differential response among variety and hybrid.

Table 36. Response of some crops (kg/ha^{-1}) in medium black soil to potassium

Crop	Variety/hybrid	Response[a] to 49.8 kg K/ha^{-1}	Response to kg/kg^{-1} K
Kharif rice	Mahsuri	280	5.6
Rabi rice	Tellahamse	409	8.2
Wheat	Kalyan sona	249	5.0
Jowar	CSH-1	134	2.7
	M-35-1	71	1.4

Source: Goswami et al. (1976); AICARP (1980–81).

[a] Response is over $N_{120}P_{26}$ kg/ha^{-1}.

Table 37. Potassium fixation capacity of vertisols of India

State	K fixed (mmol/kg^{-1})	K fixation (%)	K saturation (%)
Andhra Pradesh	37.0	9.1	1.6
Bihar	—	39.7	1.4
Gujarat	12.0	11.0	2.7
Karnataka	13.0	42.0	1.2
Madhya Pradesh	30.0	—	1.5
Maharashtra	70.0	21.3	1.0
Rajasthan	30.0	—	—
Tamilnadu	—	11.6	—
Uttar Pradesh	13.0	28.0	2.1

Source: Zende (1978).

The data shown in Table 36 indicate that rice is more responsive than wheat and that the response of rice is higher during the post-rainy season than in the rainy season. The higher response shown by rice over wheat and sorghum is possibly due to an increase in the availability of K with increase in soil moisture, since rice is grown under submerged conditions. Variations in soil moisture and temperature affect the K release and in turn the magnitude of response of sorghum and wheat to K, as shown in Table 37. These crops are generally raised during the post-rainy season where moisture content ranges from air dry to 0.15 kg/kg^{-1} with a temperature in the neighborhood of 35–40°C where the K release will be low (Goswami *et al.*, 1976; AICARP, 1980–81).

Hunsgi *et al.* (1974) observed that the response of sugarcane to K grown in black soils of Belgaum, Karnataka, was parabolic ($Y = 134.2403 + 9.529\chi - 1.4199\chi^2$) in experimental stations, while in cultivator's fields it was linear ($Y = 83.3699 + 0.1021\chi$). Such linear response could possibly be due to lower fertility status as compared to that of experimental stations.

Figure 15 shows the results of the response of sunflower (var. Morden) to added potassium in combination with N and P in medium black soils. The data indicate that application of 52 kg K/ha^{-1} along with 62.5 kg N and/or 32.7 kg P/ha^{-1} increased the sunflower seed yield by 160% and 180%, respectively, over control. An almost similar increase in oil yield was observed when K was applied along with N and/or P. The results apparently indicate that K efficiency can be increased by keeping N/P:K balance (Figure 15). Furthermore, application of K with N and P increased the potassium concentration in the leaf by 271% and 257%, respectively, over control. A positive and highly significant correlation between leaf K and seed yield ($r = 0.83$) and oil yield ($r = 0.84$) was obtained (ACBSR, 1983).

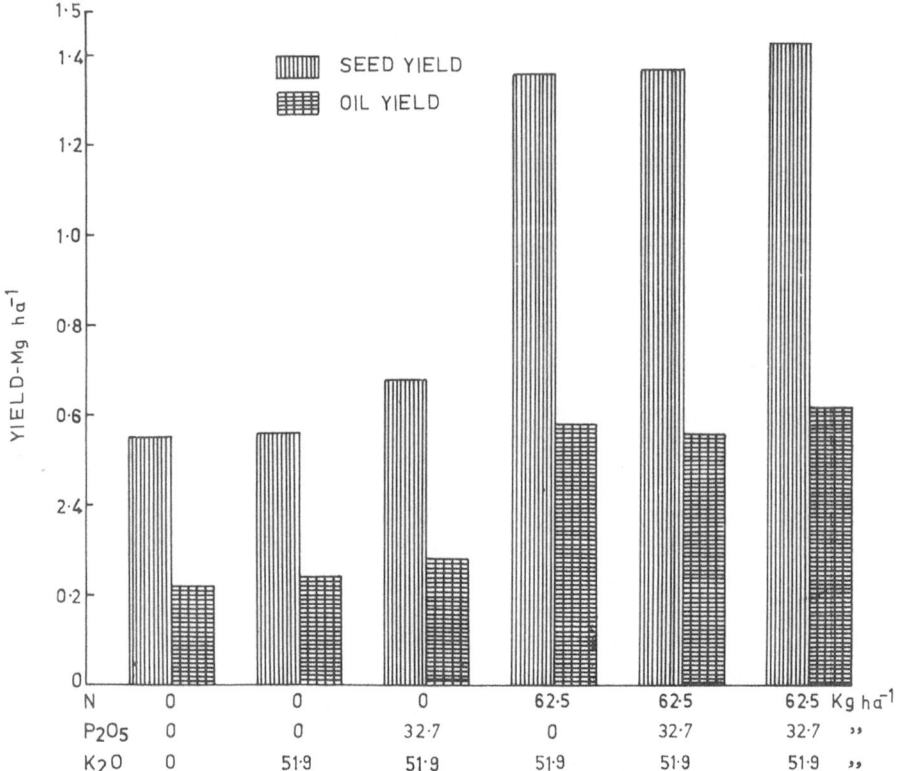

Figure 15. Response of sunflower (var. Morden) to potassium in combination with N and P (ACBSR, 1983).

E. Factors Affecting Potassium Fixation and Release

The important factors that affect K fixation in soils are pH, nature of clay minerals, CEC, and moisture regime.

1. Clay Minerals

The potassium fixation capacity of vertisols of India is shown in Table 37. The K fixation capacity ranges from 11% to 42%. K fixation capacity is high in vertisols of Karnataka and Bihar as compared with vertisols from other parts of India. The high K fixation by vertisols of Karnataka could as well be due to the presence of iron-rich beidellite (ACBSR, 1983). The data in Table 37 show that vertisols of semiarid regions fix less K than vertisols of humid regions. It is possible that soils of humid regions contain vermiculite and high amounts of organic matter.

Mitra *et al.* (1958), in their studies on K fixation under continuously moist conditions, report that vertisols of the arid and semiarid regions with mixtures of montmorillonite and kaolinite fix less than 10% K, as compared to vertisols of the humid regions containing illite besides montmorillonite (18.7%). The difference in responsiveness to added K and/or fixation of K has been attributed to the geological origin (Mitra *et al.*, 1958) of the soil besides climate and degree of K saturation. Vertisols of deccan trap origin show poor response to K and fix more of added K.

2. Moisture Regime

Hasan and Velayutham (1971), in studies on fixation of K, report greater fixation of added K (16.4%) under alternate wetting and drying cycles than under wet and dry conditions. They attribute this to the strong attraction offered by adsorbed K which precludes expansion of crystal lattice and reentrance of water. Further, they contend that soils with low K saturation fix more of K than soils with high K saturation. K fixation capacity (16.4%) obtained for black soil of Coimbatore is approximately the same as that obtained for pure bentonite by Mitra *et al.* (1958).

Mehta and Shah (1956) observed K fixation to the extent of 90% at low concentration of added K. This will be maximum in soils kept moist for 210 days. Mehta and Shah (1956) report a decrease in CEC consequent to K fixation in a vertisol from Gujarat, suggesting a possible conversion of expansible layer silicates to nonexpansible ones. Kadrekar and Kibe (1972) observed that vertisols of Maharashtra release K under moist conditions after 40 days of incubation and a slower rate of K fixation after 150 days and at higher moisture levels. In a black soil of Akola, Kharkar and Desmukh (1976) observed an increase in the availability of K with increase in soil moisture. There was a significant increase in the uptake of K and yield of cotton and sorghum with a progressive decrease in moisture.

3. Fertilizers

Results presented in Table 38 by Patil *et al.* (1976) showed an increase in the potassium fixation with increase in time of contact between the fertilizer and the soil. Potassium fixation was more in the presence of ammonium sulfate and superphosphate than with the K application individually (Table 38).

From the foregoing, it is evident that a large supply of potassium is usually present in soils. This potassium is present in the clay fraction as well as in the silt and sand fractions. Most of the potassium that is readily available to plants exists as exchangeable ions, mainly on clay mineral surfaces. Although the literature on the behavior of potassium in soils is replete, there are hardly any data where a knowledge of the mineralogical properties such as location of the negative charge, disorder within the particle, and interlayer features are combined with the release and/or fixation of K to provide a distinct picture of potassium behavior in vertisols of India.

Table 38. Effect of N and P additions of K fixation in medium black soil

Treatment	K applied (mmol/kg^{-1})	K fixed days (mmol/kg^{-1})		
		30	60	90
Control	0	1.3	0.9	0.6
KCl	200	84.2	88.3	90.9
KCl + AS	200	85.7	89.0	90.9
KCl + SP	200	88.3	91.3	93.6
KCl + AS + SP	200	88.7	91.5	93.5
K$_2$SO$_4$ + AS + SP	200	88.3	91.2	94.0
Schoenite + AS + SP	200	88.4	91.0	93.6

Source: Patil *et al*. (1976).

As, Ammonium sulfate; SP, superphosphate.

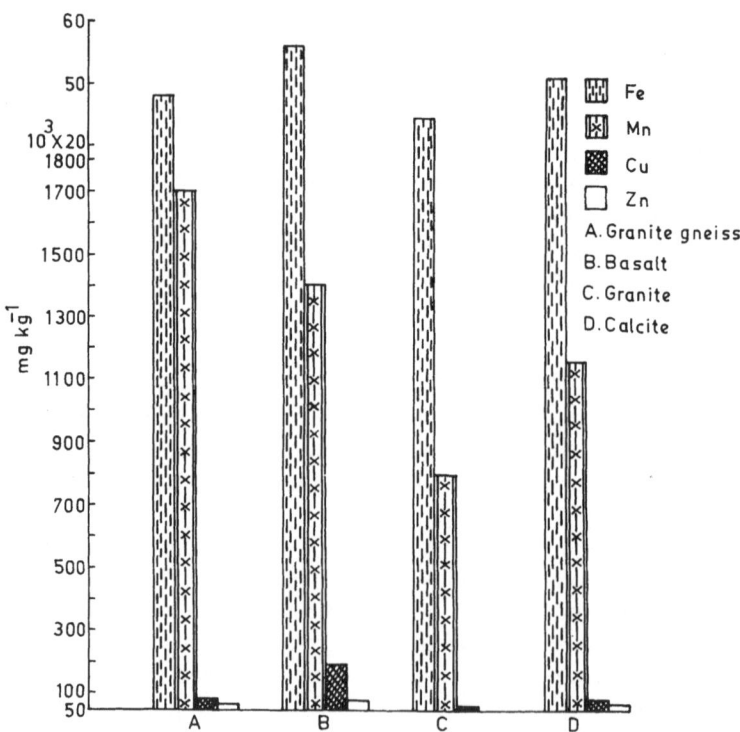

Figure 16. Total micronutrient status in vertisols of Karnataka, India, in relation to parent rock (ACBSR, 1981).

IX. Micronutrients

With the adoption of modern agricultural technology in India, deficiencies of micronutrients and spectacular responses of crops to their application are being observed. The deficiency of zinc is widespread. Iron chlorosis has been recorded in vertisols on crops like sugarcane and groundnut. In recent times, micronutrients have gained importance in crop production, and it is pertinent to assemble and present the available information on micronutrient status of vertisols and factors governing their availability as well as responses of crops to micronutrients. The information provided in this review will help in planning future research strategies for efficient micronutrient management in vertisols.

A. Distribution and Availability

Vertisols of India in general are fairly rich in total micronutrients—namely, iron, manganese, and zinc. Availability of these nutrients is less than 1% of total nutrients. The micronutrient status in vertisols of Karnataka shown in Figures 16 and 17 amply demonstrates the above point

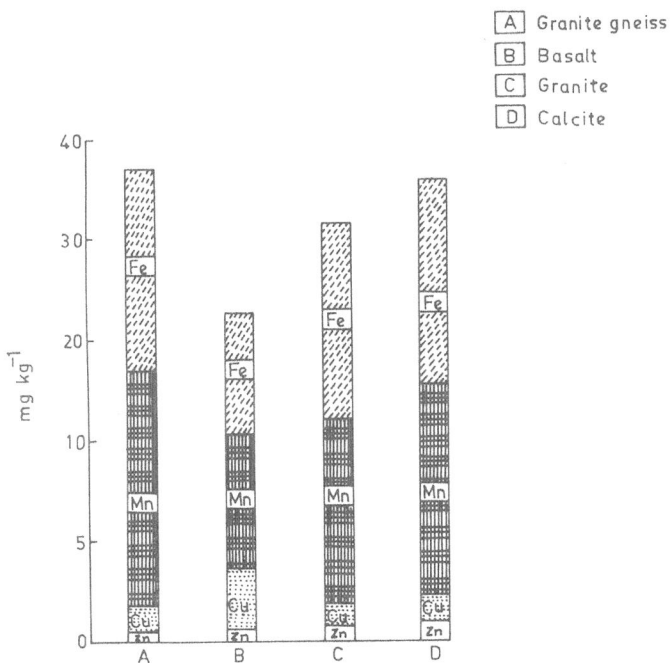

Figure 17. Available micronutrient status in vertisols of Karnataka, India, in relation to parent rock (ACBSR, 1981).

(ACBSR, 1981). Further, the data shown in Figures 16 and 17 indicate that the amounts of total and DTPA-extractable micronutrients (available) are influenced to a great extent on the nature of the parent material from which the vertisols are derived. Vertisols of granite-gneiss and calcite origin contain more of available Fe, Mn, Cu, Zn than soils derived from granite and basalt.

1. Iron and Manganese

Vertisols of India are generally rich in total iron and manganese, having values of 1–5%. Availability of iron is low because of high pH, calcareous nature, and high native Mn content.

2. Zinc

Total zinc content of vertisols is of the order of 60–90 mg/kg^{-1}, and that of available zinc is 0.5 mg/kg^{-1}. Results from AICRPM (1980) show a significant correlation between plant uptake and DTPA-extractable zinc in neutral to alkaline vertisols. On the basis of extractable zinc measurements, about 54% of the soils in Andhra Pradesh and 58% soils in Madhya Pradesh are deficient in zinc.

Tiwari et al. (1976) report an increase in zinc uptake by rice at field capacity as compared to the uptake under waterlogged conditions.

3. Copper

The total copper content of vertisols of India is 100 mg/kg^{-1}. Available copper as measured by DTPA extraction is in the range of 1–2 mg/kg^{-1}. Vertisols are well supplied with copper for plant growth, and there is hardly any report on copper deficiency in soils in spite of high pH.

4. Molybdenum

Total molybdenum content of vertisols is in the range of 1–3 mg/kg^{-1}. Molybdenum availability increases with an increase in pH and is not a problem in vertisols. However, the role of molybdenum as a plant nutrient for crops raised in vertisols is less understood.

B. Crop Response to Micronutrients

Results of yield response of wheat to zinc in vertisols of Maharashtra are shown in Figure 18 indicating that 5 kg of Zn/ha^{-1} has increased the grain yield of Sonalika wheat by 56% over no zinc treatment (Shinde et al., 1977). Results similar to this were observed in the experiments conducted by the Advance Centre for Black Soil Research, Dharwad, Karnataka, under rain-fed conditions (ACBSR, 1983). Grain yield of wheat was 1.6 t/ha^{-1} when the soil was supplied with 25 N and 5.4 kg P/ha^{-1} along with

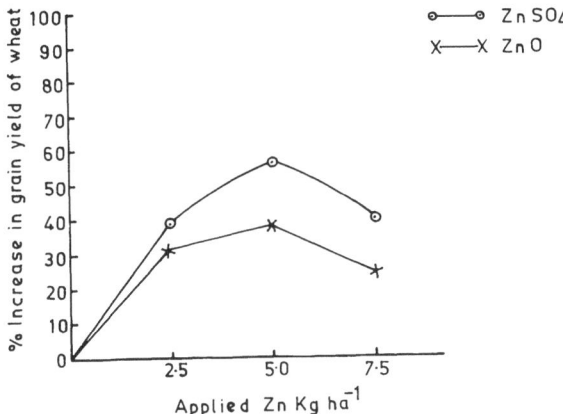

Figure 18. Wheat grain yield as affected by different forms of zinc (Shinde *et al.*, 1977).

Zn, Mn, and Fe as compared with 1.3 t/ha^{-1} under 25 N and 5.4 kg P only. This indicates that the response of wheat to micronutrients will be better when the crop is provided with a mixture of Zn, Mn, and Fe.

Nambiar and Motiramani (1981) in studies on prediction of zinc deficiency in maize report that a definite proportion of Fe/Zn in the plant tissue is necessary for optimum plant growth rather than absolute amounts. The critical Fe/Zn in maize was found to be around 6.0 and significantly correlated with dry matter yields ($r = -0.457$).

Unpublished research of the Advance Centre for Black Soil Research indicates a significant response of cowpea to applied molybdenum in irrigated vertisols of Karnataka (Figure 19). Direct application of molybdenum to cowpea crop at the rate of 0.5 kg/ha^{-1} with 10 kg N, 21.8 kg P, and 8.3 kg K/ha^{-1} showed an increase in grain yield by 0.27 t/ha^{-1} over no fertilizers. Further, the data presented in Figure 19 indicate a good residual effect of molybdenum on the post-monsoon sorghum crop (ACBSR, 1981).

C. Iron Chlorosis in Vertisols—A Rationale

The problem relating to iron availability and chlorosis in vertisols is serious and is not well understood. The problems relating to the iron and manganese availability and lime-induced chlorosis are associated with calcareous soils. At higher levels of $CaCO_3$, availability of iron and manganese decrease owing to oxidation of Fe^{2+} and Mn^{2+}. This process brings about inactivation of iron, resulting in "iron chlorosis" (Patil and Patil, 1981).

The reason for iron chlorosis in crops grown in vertisols is a result of the formation of insoluble iron phosphate on the root surface, contributing to

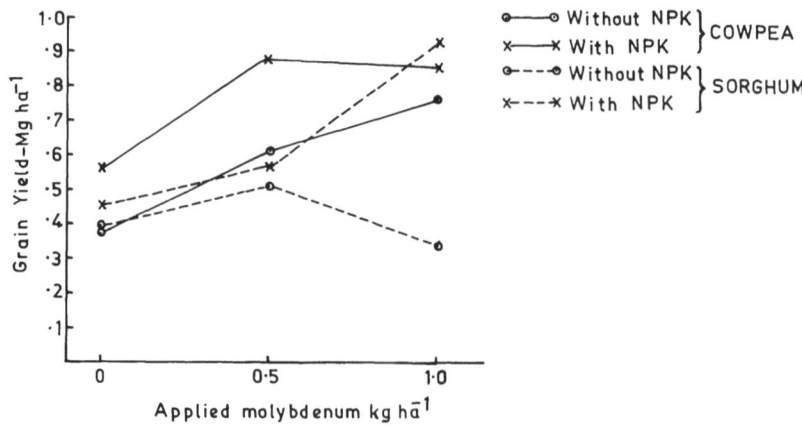

Figure 19. Response of cowpea to added molybdenum and its residual effect on rabi sorghum (ACBSR, 1981).

immobilization of Fe. During the wet phase, when soils are waterlogged either by rainfall or irrigation, iron oxides are mobilized to ferrous iron by bacteria under reducing conditions. The ferrous iron thus formed gets oxidized to Fe^{3+} at the expense of higher-valence oxides of Mn and restrict the movement of Fe from roots to other plant parts, resulting in Fe disorder and leading to the so-called iron chlorosis. This has serious implications in the iron nutrition of crop plants grown in vertisols with high native manganese. Soil application of fertilizer iron to vertisols high in native manganese will not be of any benefit and could even aggravate the problem of iron deficiency by bringing more manganese into the soil solution. Unpublished results of the Advance Centre for Black Soil Research and work carried out elsewhere suggest the use of genotypes tolerant to iron deficiency to overcome iron chlorosis in crop plants.

X. Perspectives and Research Needs

This review has shown great potential of vertisols for producing many times the crop yields currently achieved. Agricultural production capacity of vertisols can be expanded by increases both in the yields per hectare and in the area of cultivated land. This growth in resource productivity requires scientific management of soil, water, and crop through a systems approach tailored to suit the agroclimatic zones in the country.

In spite of the availability of a large body of research information on fertilizer reactions, nutrient transformation in soils, response of field crops to fertilizer elements and micronutrients, it is very difficult to predict explicit changes in soil and plant nutrient dynamics under a given set of treatments. This is mainly due to lack of adequate information on soil characteristics and/or soil types prevailing in the experimental site. In view of this, it is important to give adequate attention to soil characteristics while reporting the experimental results for a meaningful apprisal.

This review has also identified several major research gaps, such as the following:

1. An integrated water-fertility management research program with direct effect on crop yields, fertilizer, and WUE has to be developed for vertisols under irrigation.
2. Study of the changes in soil properties, both chemical and physical, with time in major soil series or soil family farming systems should be initiated. This monitoring on a limited number of benchmark locations would help in the prediction of changes in nutrient dynamics and/or soil physical degradation consequent to irrigation or a particular cropping system and correction of them before they actually happen. Studies related to soil dynamics are scanty and usually carried over a short period of time. Long-term monitoring of the changes in soil properties is needed to establish a better understanding of the effect of the intensity of cropping with low and high input systems. The efficiency of fertilizer use under different cropping systems could be answered by long-term monitoring of soil properties.
3. Nitrogen research to quantify ammonia volatilization and its control through the use of inorganic salts of Ca and K, nitrogen recovery and agronomic efficacy, nitrate nitrogen as an index of available N, and the importance of inorganic and organic forms of N in determining crop yields using labeled fertilizer ^{15}N should be the prime area of research in the years to come.
4. Phosphorus being a major constraint in vertisols for crop production, research efforts are needed on a long-term basis to identify the factors that affect the response of crop to applied P in different soil types. Research on direct, residual, and cumulative effect of P in different cropping systems should be studied on a systematic basis.
5. Identification of soils and crops that need P and K fertilization on the basis of mathematical models and modifications of soil tests for available P and K are suggested.
6. Fixation and release patterns for P and K in important soil types as related to clay minerals, ionic environment, and such factors as moisture, aeration, and soil temperature deserve attention.
7. The effect of incorporation of neemcake, crop residues like maize, and cotton stalks on the physical properties and nutrient dynamics in vertisols under irrigation should be evaluated on a long-term basis.
8. Detailed investigation into the causes and remedial measures to alleviate iron chlorosis should receive immediate attention.
9. There is an urgent need to identify crop plants that would fit soil con-

ditions as they exist rather than continually change the soil to fit the needs of the plants. This approach would be particularly important for helping to overcome some of the severe Fe, Zn, and P deficiency, salinity, and alkalinity problems in Vertisols.

10. A close liaison between soil survey and soil fertility research has to be established so that soil fertility experiments are conducted using soil family taxa as a point of reference. Soils belonging to the same family should have essentially the same management requirements, and analogous responses to soil manipulation and similar potential for crop production. Soil fertility research on this line will facilitate agrotechnology transfers to other locations where similar conditions exist.

XI. Summary

Vertisols are characterized by high percentage of clay (40–60%) dominated by chloritized iron-rich smectite. Low infiltration (nil to 0.2 m/h^{-1}), high plasticity and stickiness, low organic matter, high CEC, calcareous nature, and alkaline reaction are some of the properties associated with vertisols.

Alkali feldspars such as orthoclase, microcline, and albite are present in the sand and silt fractions in vertisols. The proportion of orthoclase to microcline could serve as a useful guide to establish the origin of the parent material and profile uniformity.

The main technologies to manage the vertisols for higher crop yields center on (1) selection of fertilizer responsive genotypes, cropping systems, and synchronization of crop growth period with moisture cycle; (2) adequate moisture conservation within the soil profile or efficient utilization of rainfall and stored moisture by adopting proper land configurations, tillage, mulches, and use of fertilizers; (3) application of chloride and nitrate salts of Ca and K with surface-placed urea to effectively reduce volatilization loss of NH_3; (4) increase of the efficiency of P fertilization by applying sufficient quantity of P to meet the P-fixing capacity of the soil and the crop requirement; and (5) promotion of nutrient recycling largely limited to utilization of crop residues such as neemcake, cotton, and maize stalks in order to recycle phosphorus, potassium, and micronutrients back into the soil.

Acknowledgments

The author is indebted to the scientists in the Advance Centre for Black Soil Research, Dharwad, India, for their helpful suggestions and for providing source material used in preparing the manuscript, and to Mr. D. Dyala Doss for collection and compilation of literature. Finally, without the patience and perseverance of Ms. S.A. Jagdale and Mr. N.S. Narthi in typing the manuscript, the completion of this paper would not have been possible.

The author is grateful to Dr. K. Krishna Murthy, Director of Research, University of Agricultural Sciences, Bangalore, and the Director General, Indian Council of Agricultural Research, New Delhi, for the encouragement given during the preparation of this paper.

References

ACBSR (Advance Centre for Black Soil Research). 1978. *Annu. Rep. 1977.* ICAR/UAS, Dharwad, India.

ACBSR (Advance Centre for Black Soil Research). 1981 *Annu. Rep. 1980.* ICAR/ UAS, Dharwad, India.

ACBSR (Advance Centre for Black Soil Research). 1982. *Annu. Rep. 1981.* ICAR/UAS, Dharwad, India.

ACBSR (Advance Centre for Black Soil Research). 1983. *Annu. Rep. 1982.* ICAR/UAS, Dharwad, India.

ACBSR (Advance Centre for Black Soil Research). 1984. *Annu. Rep. 1983.* ICAR/UAS, Dharwad, India.

ACBSR (Advance Centre for Black Soil Research). 1985. *Annu. Rep. 1984.* ICAR/UAS, Dharwad, India.

AICARP (All India Co-ordinated Agronomic Research Project). 1974. *Annu. Rep. 1973.* ICAR, New Delhi.

AICARP (All India Co-ordinated Agronomic Research Project). 1975. *Annu. Rep. 1973.* ICAR, New Delhi.

AICARP (All India Co-ordinated Agronomic Research Project). 1974–80. *Annu. Rep.* ICAR, New Delhi.

AICARP (All India Co-ordinated Agronomic Research Project). 1981. *Annu. Rep. 1980.* ICAR, New Delhi.

AICRIP (All India Co-ordinated Rice Improvement Project). 1976. *Annu. Rep. 1975.* ICAR, New Delhi.

AICRPDA (All India Co-ordinated Research Project for Dryland Agriculture). 1977. *Annu. Rep. 1976.* ICAR, New Delhi.

AICRPDA (All India Co-ordinated Research Project for Dryland Agriculture). 1982. *Annu. Rep. 1981.* ICAR, New Delhi.

AICRPM (All India Co-ordinated Research Project on Micronutrients). 1980. *Annu. Rep. 1979.* ICAR, New Delhi.

AICRPWM (All India Co-ordinated Research Project on Water Management). 1981. *Annu. Rep. 1980–81*, ICAR, New Delhi.

AICSIP (All India Co-ordinated Sorghum Improvement Project). 1984. *Annu. Rep. 1983.* ICAR, New Delhi.

Barber, S.A. 1983. Soil fertility dynamics in a systems approach. *Fertil. News* 28:43–45.

Chatterjee, R.K., and G.S. Rathore, 1976. Clay mineral composition, genesis and classification of some soils developed from basalts in Madhya Pradesh. *J. Indian Soc. Soil Sci.* 24:144–157.

Chittaranjan, S., B. Ramanath, and M.S. Rama Mohan Rao. 1980. *Mechanical*

Structures for Soil Conservation on Deep Black Soils. Ext. Bull. No. 1. Central
Soil and Water Conservation Research and Training Institute, Research Centre,
Bellary, India.

Chowdhury, S.L. 1978. All India Co-ordinated Research Project for Dryland Agriculture. *Fertil. News* 23:31–37.

Datta, N.P., and J. Venkateswarlu. 1968. Uptake of fertilizer phosphorus and nitrogen from different methods of application by low land rice growing on major
Indian soils. Trans. 9th Int. Cong. Soil Sci., Adelaide, Australia, pp. 9–18.

Doddamani, V.S. 1982. A study of forms of phosphorus, phosphate potential, Q/I
relationship and sorption in relation to phosphorus availability in selected soils of
Karnataka. Ph.D. Thesis, University of Agricultural Sciences, Bangalore, India.

Dudal, R. 1965. Dark clay soils of tropical and subtropical regions. *Agric. Dev.
Paper 83.* FAO, Rome.

El-swaify, S.A., P. Pathak, T.J. Rego, and S. Singh. 1985. Soil management for
optimized productivity under rainfed conditions in the semi-arid tropics. *Adv.
Soil Sci.* 1:1–64.

FAO (Food and Agriculture Organization). 1979. *FAO Production Year Book.*
FAO, Rome.

Fenn, L.B., and L.R. Hossner. 1985. Ammonia volatilization from ammonium or
ammonium-forming nitrogen fertilizers. *Adv. Soil Sci.* 1:124–169.

Fenn, L.B., and D.E. Kissel. 1973. Ammonia volatilization from surface applications of ammonium compounds on calcareous soils. I. General theory. *Soil Sci.
Soc. Am. Proc.* 37:855–859.

Fenn, L.B., and D.E. Kissel. 1974. Ammonia volatilization from surface application of ammonium compounds on calcareous soils. II. Effects of temperature and
rate of ammonium nitrogen application. *Soil Sci. Soc. Am. Proc.* 38:606–610.

Fenn, L.B., R.M. Taylor, and J.E. Matocha. 1981. Ammonia losses from surface
applied urea and ammonium fertilizers as influenced by rate of soluble Ca. *Soil
Sci. Soc. Am. J.* 45:883–886.

Fenn, L.B., J.E. Matocha, and E. Wu. 1982. Substitution of ammonium and potassium for added calcium in reduction of ammonia loss from surface-applied urea.
Soil Sci. Soc. Am. J. 46:771–776.

Fenn, L.B., and S. Miyamoto. 1981. Ammonia loss and associated reactions of
urea in calcareous soils. *Soil Sci. Soc. Am. J.* 45:537–540.

Finck, A., and J. Venkateswarlu. 1982. Chemical properties and fertility management of Vertisols. Vertisols and rice soils of the tropics. Proc. 12th Int. Cong.
Soil Sci., New Delhi, pp. 61–79.

Ghosh, A.B., and R. Hasan. 1980. Nitrogen fertility status of soils of India. *Fertil.
News* 25:19–24.

Ghosh, S.K., and B.S. Kapoor. 1982. Clay minerals in Indian soils. In: *Review of
Soil Research in India*, Part II. Trans. 12th Int. Cong. Soil Sci., New Delhi, pp.
703–710.

Godse, N.G., and S. Gopalkrishnappa. 1976. Studies on different forms of potassium in some black and red soils of Bellary. In: *Potassium in Soils, Crops and
Fertilizers.* Bull. No. 10. Indian Soc. Soil Sci., New Delhi, pp. 52–55.

Gopalkrishna Rao, M., M.V. Kulkarni, G.V. Havangi, B.V. Venkata Rao, and
S.V. Patil. 1975. Studies on problems of dry farming at Hagari (1906–1970).
Station Series No. 1, USA, Bangalore, India.

Goswami, N.N., and Mohinder Singh. 1976. Management of fertilizer phosphorus in cropping systems. *Fertil. News* 21:56–59.

Goswami, N.N., S.R. Bapat, C.R. Leelavathi, and R.N. Singh. 1976. Potassium deficiency in rice and wheat in relation to soil type and fertility status. In: *Potassium in Soils, Crops and Fertilizers*. Bull. No. 10. Indian Soc. Soil Sci., New Delhi, pp. 186–194.

Gupta, R.K., S.G. Sharma, G.P. Tembe, and S.S. Tomar. 1978. A new approach in farming system for problem soils of central Madhya Pradesh. *JNKVV Res. J.* 12:73–79.

Hanwante, P.R., C.S. Vaidya, and J.B. Mange. 1981. Effect of N and P rates on yield and moisture use efficiency (MUE) of safflower crop. *PKV Res. J.* 5:183–186.

Hasan, R., and M. Velayutham. 1971. Fixation of phosphate and potassium as influenced by soil type. *Madras Agric. J.* 58:613–619.

Hunsigi, G., G.V. Havanagi, and B. Puttaraj Urs. 1974. Response of sugarcane to applied fertilizers. *Potash Rev.* Subject 27, 68th Suite, pp. 1–4.

ICRISAT (International Crops Research Institute for the Semi-Arid Tropics). 1978. *Annu. Rep. 1977–78.* Patancheru P.O., A.P., India.

ICRISAT (International Crops Research Institute for the Semi-Arid Tropics). 1980. *Annu. Rep. 1979–80.* Patancheru P.O., A.P., India.

ICRISAT (International Crops Research Institute for the Semi-Arid Tropics). 1981. *Annu. Rep. 1980–81.* Patancheru P.O., A.P., India.

Jain, J.M., M.C. Sarkar, and M.L. Deori. 1981. Interaction of mechanisms of gaseous losses of nitrogen in upland soils. *J. Indian Soc. Soil Sci.* 29:97–100.

Jenny, H., and S.P. Raychaudhari. 1960. Effect of climate and cultivation on nitrogen and organic matter reserves in Indian soils. ICAR, New Delhi.

Kadrekar, S.B. 1976. Soils of Maharashtra state with reference to the forms and behavior of potassium. In: *Potassium in Soils, Crops and Fertilizers*. Bull. No. 10. Indian Soc. Soil Sci., New Delhi, pp. 33–37.

Kadrekar, S.B., and M.M Kibe. 1972. Soil potassium forms in relation to agro-climatic conditions in Maharashtra. *J. Indian Soc. Soil Sci.* 20:231–240.

Kalbande, A.R., and R. Swamynatha. 1976. Characterisation of potassium in black soils developed on different parent materials in Tungabhadra catchment. *J. Indian Soc. Soil Sci.* 24:290–296.

Kanwar, J.S., and T.J. Rego. 1983. Fertilizer use and watershed management in rainfed areas for increasing crop production. *Fertil. News* 28:33–43.

Kanwar, J.S., J. Kampen, and S.M. Virmani. 1982. Management of vertisols for maximizing crop production—ICRISAT experience. In: *Vertisols and Rice Soils of the Tropics. Symposia Papers II.* Trans. 12th Int. Cong. Soil Sci., New Delhi, pp. 94–118.

Kanwar, J.S., T.J. Rego, and N. Seetharama. 1984. Fertilizer and water-use efficiency in pearl millet and sorghum in vertisols and alfisols of semi-arid India. *Fertil News* 29:42–52.

Kharkar, P.T., and V.A. Deshmukh. 1976. Effect of moisture conditions on the availability of potassium and its uptake by the local and improved varieties of cotton and jowar. In: *Potassium in Soils, Crops and Fertilizers*. Bull. No. 10. Indian Soc. Soil Sci., New Delhi, pp. 213–218.

Kothandaraman, G.V., and K.K. Krishnamoorthy. 1979. Forms of inorganic phosphorus in Tamil Nadu soils. In: *Phosphorus in Soils, Crops and Fertilizers*. Bull. No. 12. Indian Soc. Soil Sci., New Delhi, pp. 243–248.

Krantz, B.A., J. Kampen, M.B. Russell, G.E. Thierstein, S.M. Virmani, and R.W. Willey. 1977. *The Farming Systems Research Program*. ICRISAT, Hyderabad, India.

Krantz, B.A., Kampen, J., and Russel, M.B. 1978. *Diversity of Soils in the Tropics*. Am. Soc. Agron., Madison, WI, pp. 77–95.

Krishnamoorthy, P., and S.V. Govinda Rajan. 1977. Genesis and classification of associated red and black soils under Rajolibunda diversion irrigation scheme (Andhra Pradesh). *J. Indian Soc. Soil Sci.* 25:239–246.

Krishnamoorthy, K.K., K.K. Mathan, and P.K. Mahalingam. 1976. Potassium status of Tamil Nadu soils. In: *Potassium in Soils, Crops and Fertilizers*. Bull. No. 10. Indian Soc. Soil Sci., New Delhi, pp. 38–41.

Krishna Murti, G.S.R., and K.V.S. Satyanarayana, 1969. Significance of magnesium and iron in montmorillonite formation from basic igneous rocks. *Soil Science*. 107:381–384.

Krishna Murti, G.S.R., and K.V.S. Satyanarayana, 1970. Discussion on the significance of magnesium and iron in montmorillonite formation from basic igneous rocks. *Soil Science* 110:287.

Lodha, B.K., and S.P. Seth. 1970. The relationship between different forms of potassium and particle size in different soil groups of Rajasthan. *J. Indian Soc. Soil Sci.* 18:121–128.

Magar, S.S. 1982. Soil-water management aspects related to crop production in black cotton soils (vertisols). In: *Proceedings of the Workshop in Water Resources Planning and Management*, January 22–24, 1982. WALMI, Aurangabad, Maharashtra, India, pp. 5–18.

Mahapatra, I.C., R. Prasad, K.S. Krishnan, N.N. Goswami, and S.R. Bapat. 1973. Response of rice, jowar, maize, bajra, groundnut and castor to fertilizers under rainfed conditions on farmers' fields. *Fertil. News* 18:18–28.

Mehrotra, C.L., and Gulab Singh. 1970. Forms of potassium in broad soil groups of Uttar Pradesh. *J. Indian Soc. Soil Sci.* 18:327–334.

Mehrotra, C.L., G. Singh, and R.K. Pandey. 1973. Relationship between different forms of potassium in different particle sizes in broad soil groups of Uttar Pradesh. *J. Indian Soc. Soil Sci.* 21:421–427.

Mehta, B.V. 1976. Potassium status of Gujarat soils. In: *Potassium in Soils, Crops and Fertilizers*. Bull. No. 10. Indian Soc. Soil Sci., New Delhi, pp. 25–32.

Mehta, B.V., and C.C. Shah. 1956. Potassium status of soils of western India. *Indian J. Agric. Sci.* 26:193–292.

Mehta, P.M., J.M. Patel, M.L. Patel, and A.M. Bafna. 1979. Forms of phosphorus in the profiles of deep black soils of heavy rainfall area of Gujarat. In: *Phosphorus in Soils, Crops and Fertilizers*. Bull. No. 12. Indian Soc. Soil Sci., New Delhi, pp. 249–258.

Mishra, B., B.R. Tripathi, and R.P.S. Chauhan. 1970. Studies on forms and availability of potassium in soils of Uttar Pradesh. *J. Indian Soc. Soil Sci.* 18:21–26.

Mitra, G.N., V.A.K. Sarma, and B. Rama Moorthy. 1958. Comparative studies on the potassium fixing capacities of Indian soils. *J. Indian Soc. Soil Sci.* 6:1–6.

Moraghan, J.T., T.J. Rego, R.J. Buresh, P.L.G. Vlek, J.R. Burford, S. Singh, and K.L. Sahrawat. 1984. Labeled nitrogen fertilizer research with urea in the semi-arid tropics. II. Field studies on a vertisol. *Plant Soil* 80:21–33.

More, S.D., K.G. Kachhave, and S.B. Varade. 1979. Phosphorus availability as influenced by some important properties of vertisols of Marathawada. In: *Phos-*

phorus in Soils, Crops and Fertilizers. Bull. No. 12. Indian Soc. Soil Sci., New Delhi, pp. 297–299.

More, S.D., S.B. Varade, and P.R. Bharambe. 1977. Studies on volatilization losses of ammonia from different N-carriers in calcareous vertisols. *J. Maharashtrta Agric. Univ.* 2:106–109.

Murthy, R.S. 1981. Distribution and properties of vertisols and associated soils. In: *Improving the Management of India's Deep Black Soils*. ICRISAT, Patancheru P.O., A.P., India, pp. 9–16.

Mutatkar, V.K., and S.P. Raychaudhuri. 1959. Carbon and nitrogen status of soils of arid and semi-arid regions of India. *J. Indian Soc. Soil Sci.* 7:255–262.

Nad, B.K., N.N. Goswami, and C.R. Leelavathi. 1975. Some factors influencing the phosphorus fixing capacity of India soils. *J. Indian Soc. Soil Sci.* 23:319–327.

Nagarama Murthy, G., M.A. Rahiman, and R.L. Narasimham. 1976. Some aspects of potassium availability in red and black soils. In: *Potassium in Soils, Crops and Fertilizers*. Bull. No. 10. Indian Soc. Soil Sci., New Delhi, pp. 46–51.

Nambiar, K.K.M., and D.P. Motiramani. 1981. Tissue Fe/Zn ratio as a diagnostic tool for prediction of Zn deficiency in crop plants. I. Critical Fe/Zn ratio in maize plants. *Plant Soil* 60:357–367.

Olsen, S.R., C.V. Cole, F.S. Watanabe, and L.A. Dean. 1954. Estimation of available phosphorus in soils by extraction with sodium bicarbonate. *USDA Cir. 939*, Washington.

Patil, S.P., and N.D. Patil. 1976. Effect of organic matter with nitrogenous fertilizers on leaching losses of nitrogen from black soil. *J. Maharashtra Agric. Univ.* 1:143–145.

Patil, J.D., and N.D. Patil. 1981. Effect of calcium carbonate and organic matter on the growth and concentration of iron and manganese in sorghum (*Sorghum bicolor*). *Plant Soil* 60:295–300.

Patil, A.J., S.P. Kale, and A.K. Shingte. 1976. Effect of alternate wetting and drying on fixation of potassium. In: *Potassium in Soils, Crops and Fertilizers*. Bull. No. 10. Indian Soc. Soil Sci., New Delhi, pp. 132–137.

Patil, N.D., N.K. Umrani, S.A. Shende, B.S. Manke, S.P. Kale, and A.K. Shingte. 1981. Improved crop production technology for drought prone areas of Maharashtra. *Tech. Bull. Mahatma Phule Agricultural University*, Solapur, Maharashtra, India.

Patil, B.B., N.K. Umrani, and S.H. Shinde. 1982. *Role of Legumes in Nitrogen Economy of Crops*. Mahatma Phule Agricultural University, Rahuri, Maharashtra, India.

Prasad, R., and B.V. Subbaiah. 1982. Nitrogen the key plant nutrient in Indian agriculture. *Fertil. News* 27:27–42.

Puranik, R.B., D.K. Ballal, and N.K. Barde. 1978. Studies on nitrogen forms as affected by long-term manuring and fertilization in vertisols. *J. Indian Soc. Soil Sci.* 26:169–172.

Ramanathan, K.M. 1975. Potassium release characteristics of certain soils of Tamil Nadu. *Madras Agric. J.* 62:1–9.

Randhawa, N.S., and M.S. Rama Mohan Rao. 1981. Management of deep black soils for improving production levels of cereals, oilseeds and pulses in the semi-arid region. In: *Improving the Management of India's Deep Black Soils*. ICRISAT, Patancheru. P.O., A.P., India, pp. 67–79.

Raychaudhuri, S.P., B.B. Roy, S.P. Gupta, and M.L. Dewan. 1963. *Black Soils of*

India. National Institute of Sciences of India, New Delhi, pp. 1–8.

Rego, T.J., J.T. Morghan, and Sardar Singh. 1982. Some aspects of soil nitrogen relating to double cropping of "deep" vertisols in the SAT. *Trans. 12th Int. Cong. Soil Sci.* Vol. 6, Jan. 8–16, 1982, New Delhi, p. 486.

Sahrawat, K.L. 1977. EDTA extractable P in soils as related to available and inorganic P forms. *Comm. Soil Sci. Plant Anal.* 8:281–287.

Sahrawat, K.L. 1979. Nitrogen losses in rice soils. *Fertil. News* 24:38–48.

Sahrawat, K.L., and J.R. Burford. 1982. Modification of the alkaline permanganate method for assessing the availability of soil nitrogen in upland soils. *Soil Sci.* 133:53–57.

Shinde, P.H., J.M. Khilari, S.P. Kale, and G.S. Khanvilkar. 1977. Effect of graded levels of zinc compounds applied through two sources on yield of wheat. *J. Maharashtra Agric. Univ.* 2:214–216.

Shukla, G.C., and M. Singh. 1968. A new photo-electric colorimetric method for the estimation of nitrate nitrogen in soils. *J. Indian Soc. Soil Sci.* 16:77–81.

Singhania, R.A., and N.N. Goswami. 1978. Transformation of applied phosphorus in soils under rice-wheat cropping sequence. *Plant Soil* 50:527–535.

Singh, G., and G.S.R. Krishna Murti, 1974. Mineralogy of a few basaltic soils of Madhya Pradesh. In: *Proceedings of the Indian National Science Academy*. 40B:338–345.

Sinha, S.K., and M.S. Swaminathan. 1979. The absolute maximum food production potential in India—an estimate. *Curr. Sci.* 48:425–429.

Soil Survey Staff. 1975. Soil taxonomy. A basic system of soil classification for making and interpreting soil surveys. In: *USDA Hand Book 436*. USDA, Washington.

Subbiah, B.V., and M.S. Sachdev. 1981. Fate of fertilizer nitrogen in soil-plant system—quantitative approach using ^{15}N as a tracer. *Fertil. News* 26:31–35.

Tandon, H.L.S. 1974. Dynamics of fertilizer N in Indian soils. I. Usage, transportation and crop removal of N. *Fertil. News* 19:3–11.

Tiwari, K.N., A.N. Pathak, and R.L. Upadhyay. 1976. Studies on Fe and Zn nutrition of rice at varying moisture regimes in black clay soil of Uttar Pradesh. *J. Indian Soc. Soil Sci.* 24:303–307.

Umrani, N.K., and P.G. Bhoi. 1980. *Fertil. News* 25:18–23.

Umrani, N.K., and N.D. Patil. 1983. Advances in fertilizer management for rainfed sorghum. *Fertil. News* 28:57–61.

Verma, O.P., and G.P. Verma. 1968. Studies on available potassium in soils of Madhya Pradesh. *J. Indian Soc. Soil Sci.* 16:61–64.

Virmani, S.M., M.V.K. Sivakumar, and S.J. Reddy. 1978. *ICRISAT Res. Rep. 1*. Hyderadab, India, 128 pp.

Zalawadia, N.M., and M.S. Patel. 1983. Growth response and phosphorus uptake by groundnut in calcareous soil in relation to applied phosphorus under varying soil moisture conditions. *J. Indian Soc. Soil Sci.* 31:486–490.

Zende, G.K. 1978. Potassium dynamics in black soils. In: *Potassium in Soils and Crops*. Potash Research Institute of India, New Delhi, pp. 51–68.

Index